This is an uncensored fascimile of the wartime (1942) edition of this book, with the original blemishes and printing anomalies.

Copyright © 2020 Tank Archives Press

All rights reserved. No part of this publication may be reproduced or stored in a retrieval system or transmitted, in any form or by any means, electronic, mechanical, photocopying, recording or otherwise, without prior permission in writing from the Tank Archives Press.

To request permission, write to the editor at the address below.

First published in 2020

ISBN: 978-1-951171-01-8

Published by:
Tank Archives Press

PO Box 181806
Coronada, CA 92178, USA

Editor: Bruce Oliver Newsome, PhD.

DISTRIBUTION

R.A.C	Scale C
R.A. (A. Tk)	Scale C
Reconnaissance Corps	Scale C
R.A.C., O.C.T. Us	Scale II

CONTENTS

	PAGE
Part I - Intelligence notes	1
Part II - Recognition and data of A.F.V.s.	38
Section 1 - Characteristics of British Design	39
Section 2 - Characteristics of enemy tanks	42
Section 3 - Drawings and salient features of British A.F.Vs.	49
Section 4 - Drawings and specifications of enemy A.F.Vs.	68
Section 5 - Vulnerability of A.F.Vs.	106

PART I.—INTELLIGENCE NOTES

GERMAN ARMY

(NOTES FOR USE IN THE FIELD)

No training is complete without a knowledge of the enemy. Short of knowing his organization and methods by heart, the fighting soldier needs a ready means of reference so that he can recognize and sum up his opponent and report accurately what he has seen. This book is designed to meet that need. It is not intended to replace " NOTES ON THE GERMAN ARMY—WAR " or other War Office publications; it is an attempt to condense the essential information from those publications into pocket notes which can conveniently be carried in mobile warfare.

The book is primarily intended for armoured formations in the field; for this reason special attention has been paid to tanks and A.-Tk. weapons.

The evacuation of prisoners presents special difficulties to armoured formations. Accurate identification by forward troops is therefore of great importance. Preliminary interrogation must not however be attempted except by personnel who have attended at least a local interrogation course. In a rapidly moving battle even the briefest preliminary interrogation may be impossible, and I.Os. may have to rely upon a rapid searching of documents to secure identifications. The section in this book on identification has been drawn up with these considerations in mind.

Finally this book is primarily intended for use in active operations, but its full value will be lost if it is not tried out during training.

INTENTIONALLY BLANK

INF. DIV.

(Fighting Strength: Approx. 14,000 men)

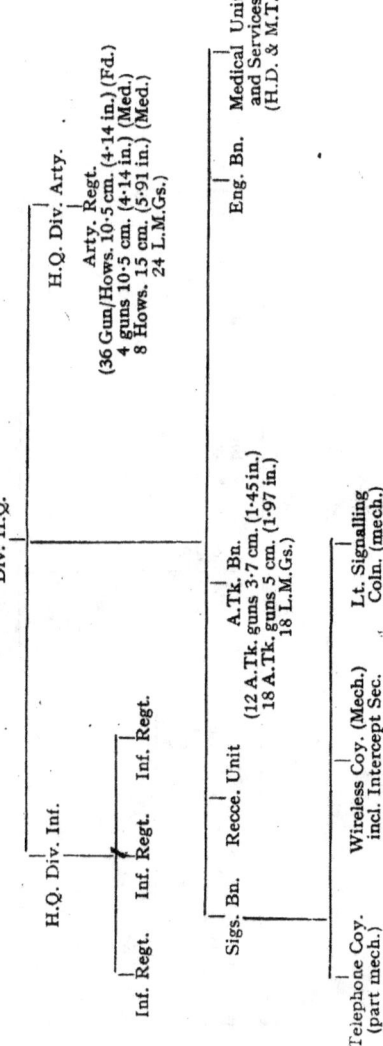

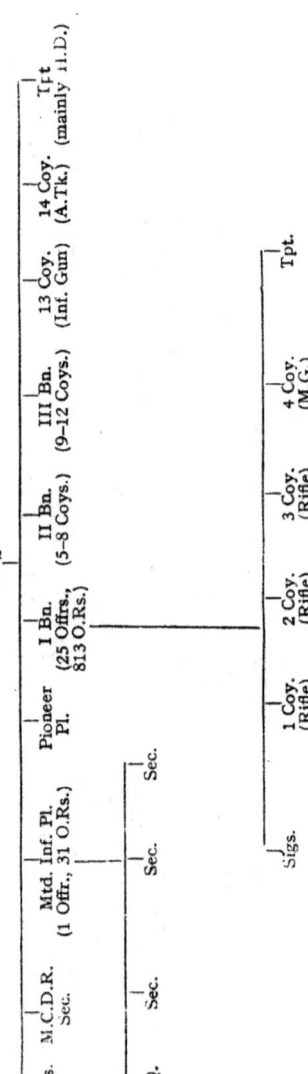

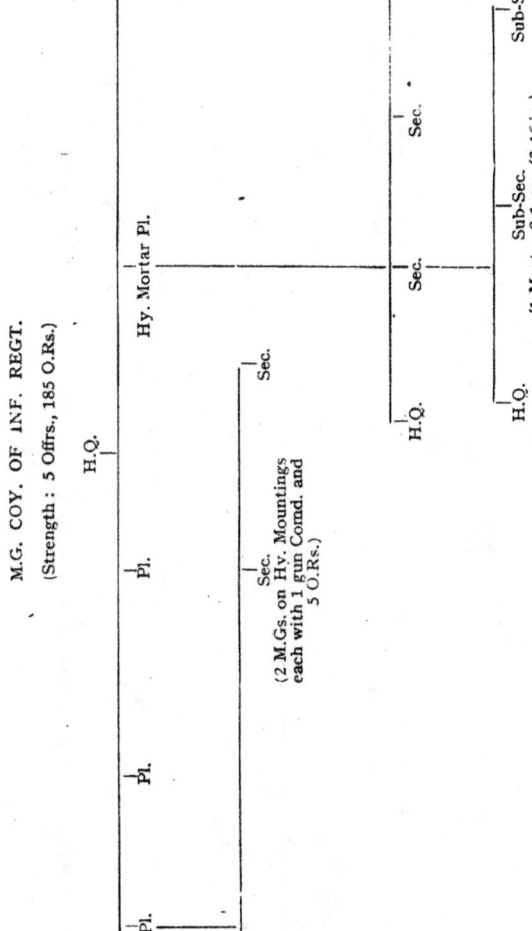

13 (INF. GUN) COY. OF INF. REGT.

(Horse-drawn. Manned by Inf. Personnel)

(Strength: 5 Offrs., 185 O.Rs.)

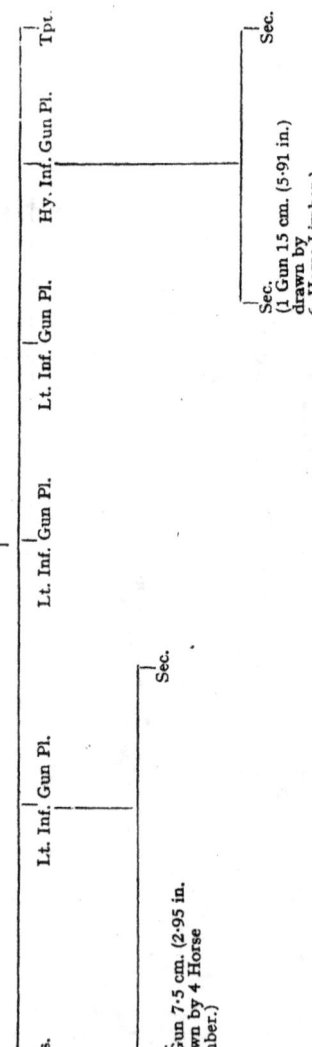

```
                                    H.Q.
  |         |           |              |             |           |
Sigs.  Lt. Inf.    Lt. Inf.      Lt. Inf.       Hy. Inf.       Tpt.
       Gun Pl.     Gun Pl.       Gun Pl.        Gun Pl.
          |                                        |
         Sec.                                     Sec.
  (1 Gun 7·5 cm. (2·95 in.                (1 Gun 15 cm. (5·91 in.)
   drawn by 4 Horse                        drawn by
   Limber.)                                6 Horse Limber.)
```

NOTE.—Assault guns, 7·5 cm. (2·95 in.) mounted on Pz. Kw. III chassis and 15 cm. (5·91 in.) on Pz. Kw. I, may be used to reinforce this Coy.

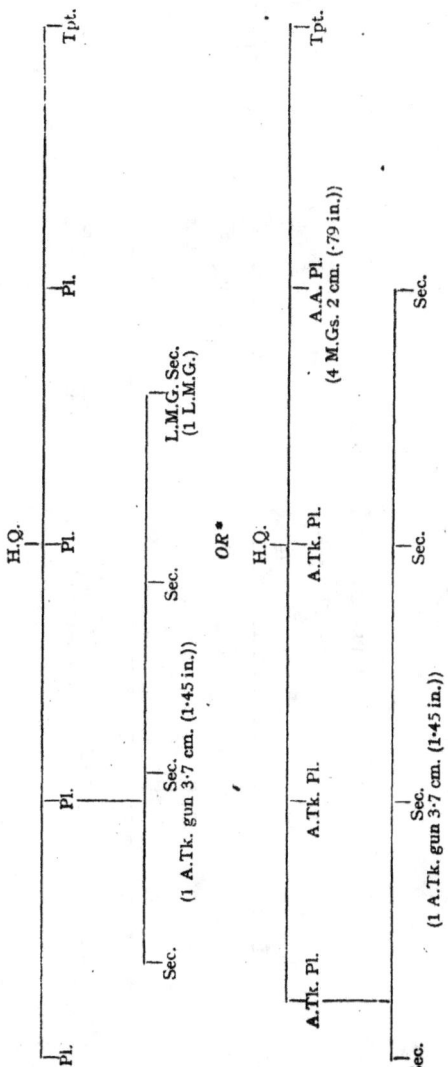

RIFLE COY. OF INF. REGT.
(Strength: 4 Offrs., 183 O.Rs.)

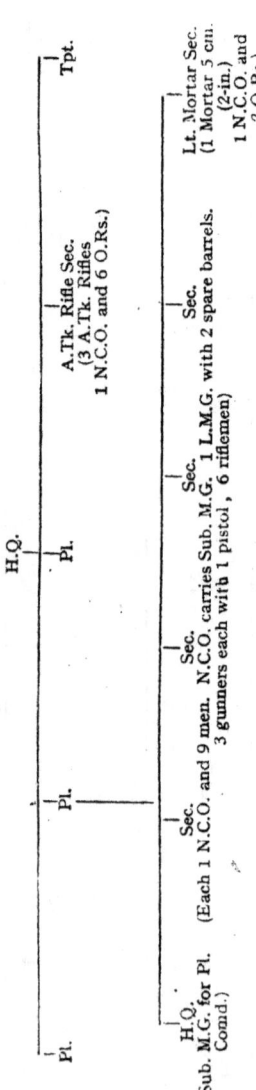

H.Q.

Pl. — Pl. — Pl. — Tpt.

H.Q. (Sub. M.G. for Pl. Comd.)

Sec. — Sec. — Sec.
(Each 1 N.C.O. and 9 men. N.C.O. carries Sub. M.G. 1 L.M.G. with 2 spare barrels. 3 gunners each with 1 pistol, 6 riflemen)

A.Tk. Rifle Sec.
(3 A.Tk. Rifles
1 N.C.O. and 6 O.Rs.)

Lt. Mortar Sec.
(1 Mortar 5 cm.
(2-in.)
1 N.C.O. and
2 O.Rs.)

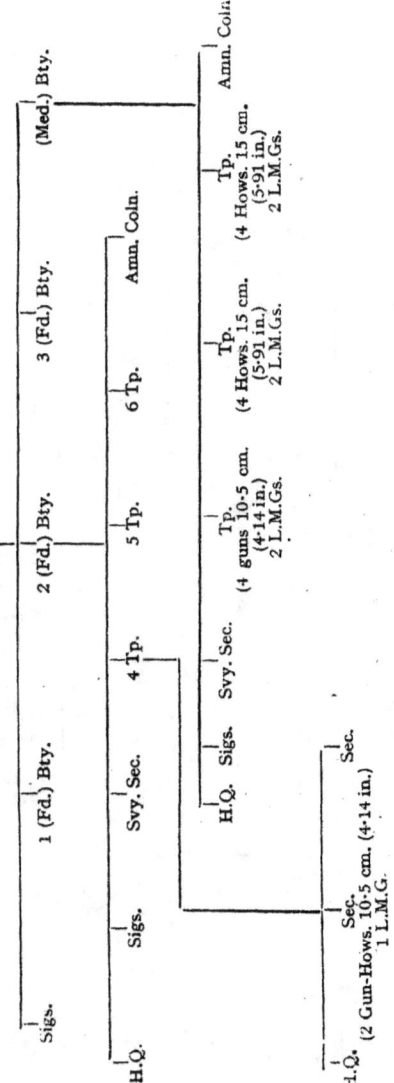

INF. DIV. RECCE. UNIT

(Strength: approx. 15 Offrs., 560 O.Rs.)

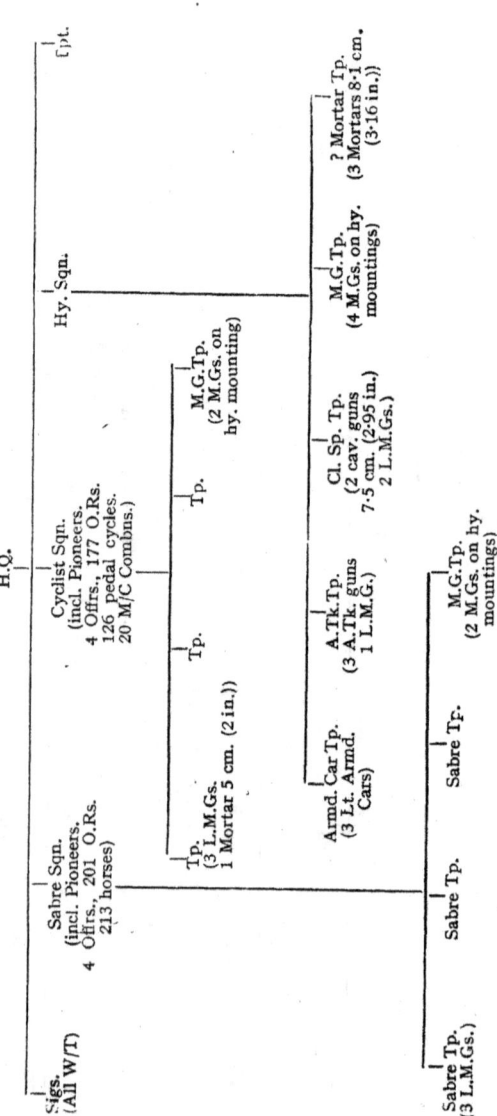

ANTI-TANK BATTALION

Headquarters
Signal Section (2 M.Cs, 10 wireless ones)
Medical Platoon

- **1 Company**
  - Coy. H.Q.
  - Coy. Comd.
  - C.S.M.
  - 4 Motor-cyclists
  - Orderly
  - One office truck

- **2 Company**

- **3 Company**
  - 1 Platoon (Light)
  - 2 Platoon (Med.)
  - 3 Platoon (Med.)
    - Pl. H.Q. (as for Light Platoon + 2 tractors with trailers, 2 drivers)
    - Sec. — 5 cm. (1·97 in.) gun, crew of 7 men, 1 truck, 1 driver
    - Sec.
    - Sec.
  - Pl. H.Q.
  - Pl. Comd.
  - Sjt.
  - 4 Motor-cyclists
  - 2 O.Rs.
  - 2 L.M.Gs.
  - Sec. — 3·7 cm. (1·45 in.) gun, each a crew of 4 men, 1 truck, 1 driver
  - Sec.
  - Sec.

- **Light Ammunition Column** (10 Lorries, 1 M.C.)
  - Coy. Transport
  - P.O.L. lorries (1,320 galls.)
  - Baggage and supply truck
  - Workshop lorry
  - Ammunition lorry
  - Field kitchen
  - 2 M.Cs.

- **Light Sup. Column** (20 Lorries, 2 M.Cs.)

Strength: 25 officers, 574 men

Armament:
3·7 cm. (1·45-in.) A.Tk. guns ... 12
5 cm. (1·97-in.) A.Tk. guns ... 18
L.M.G. 18

Transport (approx.)
Motor-cycles 64
Tractors 44
Other M.T. vehicles (approx.) ... 68

MOT. A.A. BN.
(Mech.)

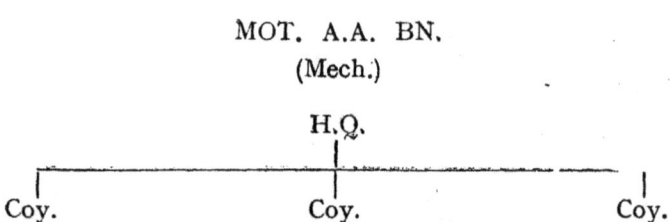

These Bns. may be equipped with any of the following weapons :—

        M.Gs. 1·5 cm. (·59 in.)
        A.A./A.Tk. M.Gs. 2 cm. (·79 in.)
        4 barrelled A.A./A.Tk. M.Gs. 2 cm. (·79 in.)
        A.A./A.Tk. guns 3·7 cm. (1·45 in.)

The total number of guns in a Bn. is believed to be 36, but the number of each calibre issued is not known and may vary.

During 1940 Bns. generally towed their M.Gs. behind tractors ; A.A. guns are now frequently mounted on S.P. mountings in the form of semi-tracked carriers.

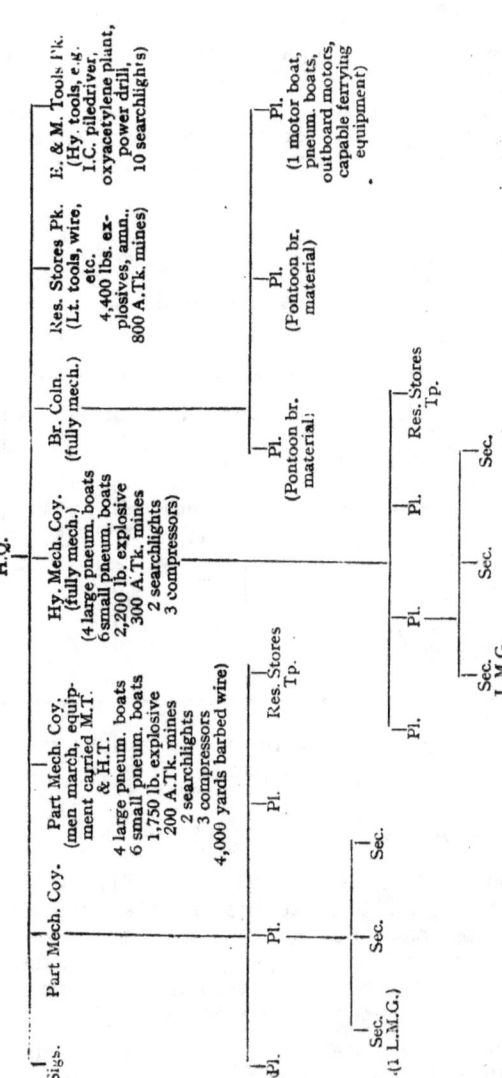

NOTES ON ENGINEERS

General

Besides being responsible for bridging, ferrying, defensive demolitions, laying of minefields and other normal engineer duties, the German sapper is an assault engineer. He leads the attacking party across rivers or over obstacles making use of explosives, flame throwers and other special equipment described below.

Bridging

Br. Coln. transports and maintains the br. equipment and the brs. are erected by the coys.

Equipment	Load Capacity in Tons	Length in Feet
TYPE B PONTOON AND TRESTLE.		
Standard Assy. Med.	9	252–270
Standard Assy. Hy.	18 (nominal rating. Will probably take about 27).	165–177
Special Assy. Lt.	4	402–432
Special Assy. Med.	8	234–264
TYPE K.		
Box Girder Bridge	9	80
(4-girder construction).	19	64
	25	48

All eng. coys. carry timber of various sizes to build brs. up to 8 tons carrying capacity.

Tank ferries

None of the brs. carried in the div. br. coln. can carry more than 22 tons at the outside. For carrying across wet gaps loads heavier than can be taken by the "B" equipment, rafts are constructed consisting of the "K" box girder equipment supported on double piers made out of "B" equipment pontoons.

Thus a good water obstacle will hold up hy. tanks for considerably longer than lt. or lt. med. tanks.

The following boats are available :—

Type	By whom carried	No. carried	Dimensions Length	Width	Capacity
Pneumatic boats large.	Coy. (mech or part mech.).	4	18 ft.	6 ft.	2¾ tons (L.M.G. ½ Sec.) (M.G. Sec.).
Pneumatic boats large.	Br. Coln.	24	18 ft.	6 ft.	2¾ tons (A.Tk.Gun).
Pneumatic boats small.	Coy. (mech. or part mech.).	6	10 ft.	4 ft.	5½ cwt. 3 men.
Pneumatic boats small.	Br. Coln.	22	10 ft.	4 ft.	5½ cwt. 3 men.
Assault boats (with outboard motors 15-20 knots).	Br. Coln.	8	?	?	18 men.
Collapsible canoes.	Br. Coln.	No details available; used for recce.; probably holds 2 men.			

The **motor boat** carried by the br. coln. is used only for assisting pontoon br. operations **not** for tpt. of tps.

Pontoons are also propelled by outboard motors.

Cable ferrying equipment consists of 2 tripods about 10 ft. high, a drum of steel wire rope, etc. Tripods must not be more than 200 yards apart; easy to erect.

A.-Tk. mines

1. **Lt. A.-Tk. mine (teller mine).**—Weight, 19 lb.; diameter, 1 ft.; height, 4 in.; filling, 11 lb. of T.N.T. When exposed looks like inverted soup plate, but usually buried and camouflaged. Sometimes arranged to detonate if moved.

2. **Hy. A.-Tk. (box) mine.**—Cast iron box, 17 in. by 16 in. by 10½ in.; filling, 37 lb. T.N.T. Not likely to be met with in the field. Used only in well prepared positions. If met, should only be neutralized by R.E. as anti-handling booby traps are incorporated. This type is not issued to ordinary field engineer units.

Assault engineers

Field engineers are armed as infantry and receive special training in assault operations. They lead the attack in river crossings, also in assaults through minefields, wire entanglements and other anti-tank and anti-personnel obstacles, and in the final reduction of pillboxes, these operations being carried out, where possible, under cover of darkness or smoke. Special equipment used in these assault operations include grenades, flame throwers, bangalore torpedoes, Cordtex nets, pole charges, etc.

Flame throwers.—The portable type flame thrower now in use consists of a large inverted steel bottle carried on the man's back containing compressed nitrogen and oil and a projector tube for directing the flame. The range is from 16–20 yards with a total duration of continuous jet of about 10 seconds. The operator is most vulnerable since a bullet or splinter in the cylinder when full will cause the latter to explode.

The bangalore torpedo consists of 2 in. steel piping in 6 ft. units arranged to join together by means of spigot and socket and containing T.N.T. It is pushed under and through a barbed wire entanglement and will blow a gap about 6 yards wide. Other improvised torpedoes are made up of grenades or T.N.T. slabs tied to long boards for use in the same manner.

Cordtex nets are nets of 6 in. mesh made of instantaneous fuze carried rolled up in units 50 ft. long by 8 ft. broad. They are laid over anti-tank minefields and detonated, thus blowing up and destroying the mines.

Pole charges are slabs of high explosive, or bundles of grenades, carried on the end of a long pole, and fired by safety or electric fuze. Their purpose is to blow in loopholes and embrasures which cannot be reached by hand.

17

MOTORIZED DIV.

(Fighting strength: approx. 12,000 men)

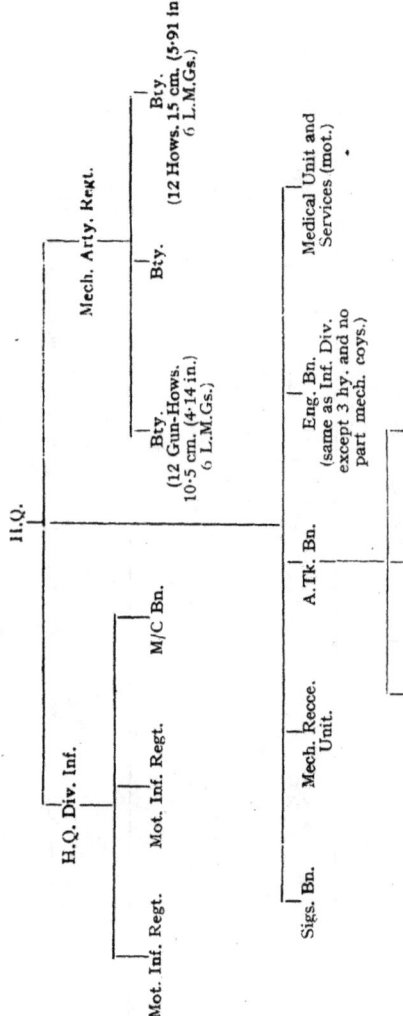

NOTES.—(i.) A.Tk. Bn. has same organization as that in an Inf. Div.

(ii.) Mech. recce. unit is organized in the same way as that in an armd. div., except that it has only one armd. car sqn.

MOT. INF. REGT. IN MOT. DIV.

(Strength: approx. 3,100 men)

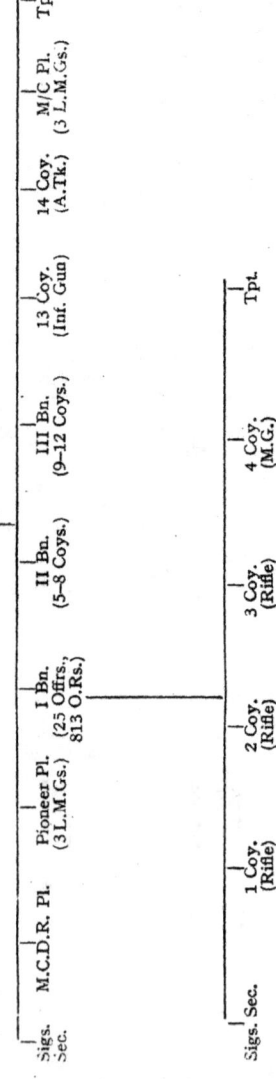

Note.—The regt. sigs. sec., M.C.D.R. pl., and pioneer pl. are grouped together as a H.Q. coy.

MOUNTAIN DIV.
(Fighting strength: approx. 11,500 men)
H.Q.

- H.Q. Div. Inf.
  - Mtn. Regt.
    - Bn.
      - Rifle Coy.
      - Rifle Coy.
      - Rifle Coy.
        - each 18 L.M.Gs. 3 5 cm. (2 in.) Mortars
    - Bn.
  - Mtn. Regt.
    - Bn.
    - 16 Coy. (A.Tk.) 12 3·7 cm. (1·45 in.) A.Tk. Guns
    - M.G. Coy. 8 Hy. M.Gs. 6 8·1 cm. (3·16 in.) Mortars
    - Hy. Coy.
      - Eng. Pl. 4 L.M.Gs.
      - Inf. Gun Sec. 2 7·5 cm. (2·95 in.) Inf. Guns
- Arty. Regt.
  - Sigs.
  - Bty.
  - Bty.
    - Tp. (4 guns 7·5 cm. (2·95 in.))
    - Tp.
  - Bty. 12 10·5 cm. (4·14 in.) Mtn. Hows.
- Recce. Unit (Part mech.)
- Sigs. Bn.
- A.Tk. Bn.
- Eng. Bn.
- Medical Unit and services

NOTES.—1. Recce. unit is organized in the same way as that in inf. div.
2. Transport of a mtn. div. is mainly on a pack basis, but includes a proportion of M.T.
3. Some mtn. rifle regts. are known to include personnel who have received training as parachutists.

ARMD. DIVS.   SECRET.

1. Normal armoured divisions

The composition of a normal armoured division is as follows :—

    H.Q.
    Div. Recce. Unit (*see* page 24).
    Div. Signals.
    Tank Regiment (*see* page 21).
    Lorried infantry brigade (of two lorried inf. regts. and one M/C bn.) (*see* pages 22–23).
    Div. arty. regt. (*see* page 25).
    Div. A.-Tk. bn. (*see* page 11).
    Div. eng. bn. (*see* page 26).
    Div. med. unit.
    Services.
    A.A. bty. (G.A.F., attached).

2. Heavy armoured divisions

Reports have been received from time to time that heavy armoured divisions are being organized. There is, however, no confirmation that heavy armoured divisions exist, and no heavy tanks have as yet been identified with German units.

3. Light armoured divisions

Reports have also been received of the organization of light armd. divs. Only one light armoured division has, however, so far been identified. This was the 5th Light (Colonial) Armoured Division in North Africa, which was, however, subsequently reorganized as an armoured division of normal type.

21

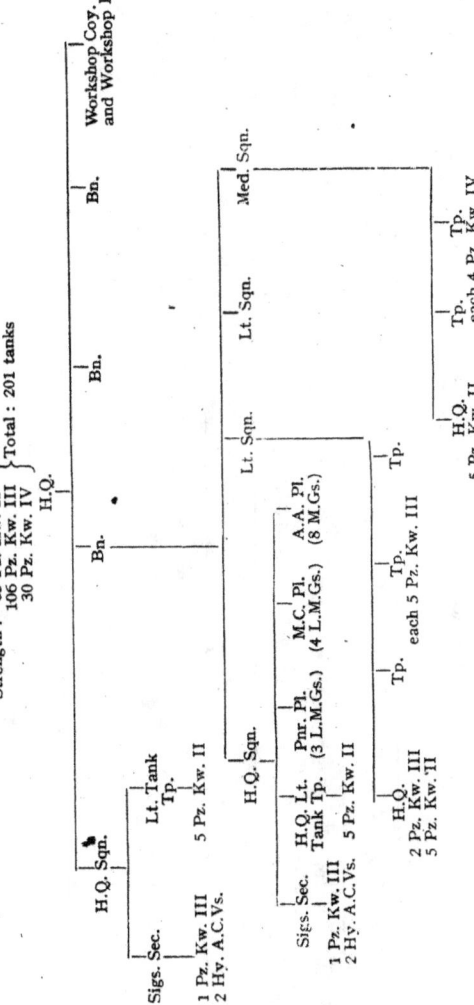

NOTES.—1. Attention is drawn to the high proportion of light medium tanks in the tank regt.
2. Every tank battalion is intended to have a reserve tank echelon of 6 tanks (2 Pz. Kw. II, 3 Pz. Kw. III and 1 Pz. Kw. IV). Each tank bn. also has a lt. sqn. which is a reserve and advanced training sqn. (numbered 3, 7, 11 as the case may be). The total tank strength (including reserves) of the tank regiment is thus 285 tanks.

LORRIED INFANTRY REGT.

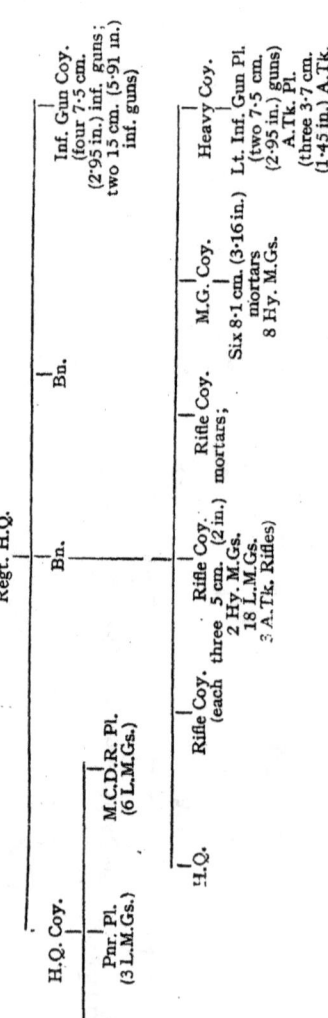

Regt. H.Q.

- H.Q. Coy.
  - Sig. Pl.
  - Pnr. Pl. (3 L.M.Gs.)
  - M.C.D.R. Pl. (6 L.M.Gs.)
- Bn.
  - H.Q.
  - Rifle Coy. (each three 2 Hy. M.Gs. 18 L.M.Gs. 3 A.Tk. Rifles)
  - Rifle Coy. 5 cm. (2 in.) mortars;
  - Rifle Coy.
  - M.G. Coy. Six 8·1 cm. (3·16 in.) mortars 8 Hy. M.Gs.
- Bn.
  - Inf. Gun Coy. (four 7·5 cm. (2·95 in.) inf. guns; two 15 cm. (5·91 in.) inf. guns)
  - Heavy Coy.
  - Lt. Inf. Gun Pl. (two 7·5 cm. (2·95 in.) guns)
  - A.Tk. Pl. (three 3·7 cm. (1·45 in.) A.Tk. guns and 1 L.M.G.)
  - Pnr. Pl. (3 L.M.Gs.)

23

M/C BN.

- H.Q.
- M/C Rifle Coy. ── M/C Rifle Coy. ── M/C Rifle Coy. (Each three 5-cm. (2-in.) mortars 2 Hy. M.Gs. 18 L.M.Gs. 3 A.Tk. Rifles)
- M.G. Coy. — Six 8·1-cm. (3·16-in.) mortars 8 Hy. M.Gs.
- Heavy Coy. — Lt. Inf. Gun Pl. (two 7·5-cm. (2·95-in.) inf. guns) A.Tk. Pl. (three 3·7-cm. (1·45-in.) A.Tk. guns and 1 L.M.G.) Pi. Pl. (3 L.M.Gs.)

MECH. RECCE. UNIT

Strength: 27 Offrs., 760 O.Rs.

H.Q. & Sig. Tp.

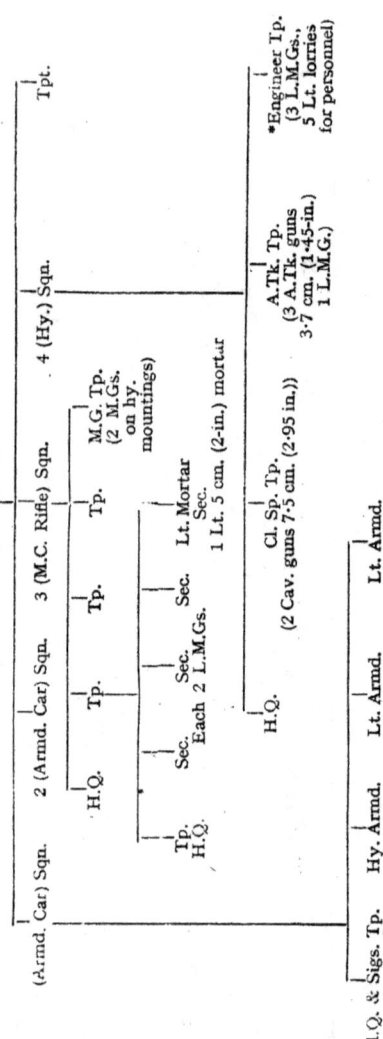

NOTE.—*Engs. can build 5-ton 36-ft. br. and two 2-ton rafts or one 4-ton raft.

DIV. ARTY. REGT. OF ARMD. DIV.
(fully mech.)

Organization consists of two field Btys. each of 12 10·5 cm (4·14 in) gun-hows. and a third medium Bty. of 12 15 cm. (5·91 in.) hows. All Btys. have 6 L.M.Gs. for local protection.

An arty. survey troop (including flash-spotting and sound-ranging elements) is attached to Regt. H Q.

26

ENG. BN. OF ARMD. DIV.

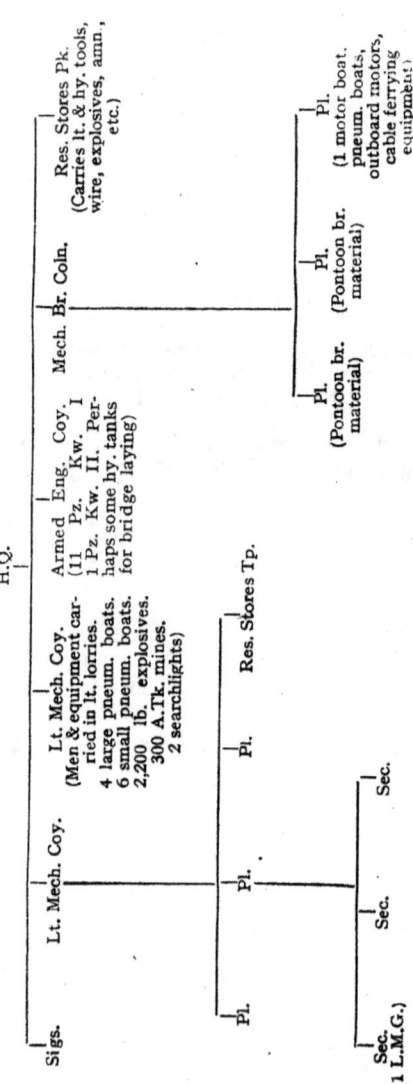

H.Q.

- Sigs.
- Lt. Mech. Coy.
  - Pl.
  - Pl.
    - Sec.
    - Sec.
      - Sec. (1 L.M.G.)
- Lt. Mech. Coy. (Men & equipment carried in lt. lorries. 4 large pneum. boats. 6 small pneum. boats. 2,200 lb. explosives. 300 A.Tk. mines. 2 searchlights)
  - Res. Stores Tp.
  - Pl.
- Armed Eng. Coy. (11 Pz. Kw. I 1 Pz. Kw. II. Perhaps some hy. tanks for bridge laying)
- Mech. Br. Coln.
  - Pl. (Pontoon br. material)
  - Pl. (Pontoon br. material)
  - Pl. (1 motor boat. pneum. boats, outboard motors, cable ferrying equipment)
- Res. Stores Pk. (Carries lt. & hy. tools, wire, explosives, amn, etc.)

ARMOURED DIVISION—SUMMARY OF A.F.Vs. AND WEAPONS

Unit	Pz. Kw. IV	Pz. Kw. III	Pz. Kw. II	Heavy Armd. Cars	Lt. Armd. Cars	L. MGs.	Hy. MGs.	A.Tk. Rifles	2 cm. (·79-in.) super heavy M.Gs.	3.7 cm. (1·45-in.) guns	5 cm. (1·97-in.) A.Tk. guns	7.5 cm. (2·95-in.) guns	15 cm. (5·91-in.) inf. guns	10·5 cm. (4·14-in.) gun/hows.	15 cm. (5·91-in.) hows.	5 cm. (2-in.) mortars	8·1 cm. (3·16-in.) mortars
Div. H.Q.	—	—	—	—	—	—	2	—	—	—	—	—	—	—	—	—	—
Recce. Unit	—	—	—	12	36	72	2	—	20	3	—	2	—	—	—	3	—
Div. Sigs. Bn.	—	—	—	—	—	17	—	—	—	—	—	—	—	—	—	—	—
Tank Regt.	30	106	65	—	—	362	24	—	65	—	106	30	—	—	—	—	—
A.Tk. Bn.	—	—	—	—	—	18	—	—	—	12	18	—	—	—	—	—	—
Lorried Inf. Bde.	—	—	—	—	—	308	70	45	—	15	—	18	4	—	—	45	30
Div. Arty.	—	—	—	—	—	18	—	—	—	—	—	—	—	24	12	—	—
Engineer Bn.	—	—	—	—	—	37	—	—	1	—	—	—	—	—	—	—	—
Total ...	30	106	65	12	36	832	98	45	86	30	124	50	4	24	12	48	30

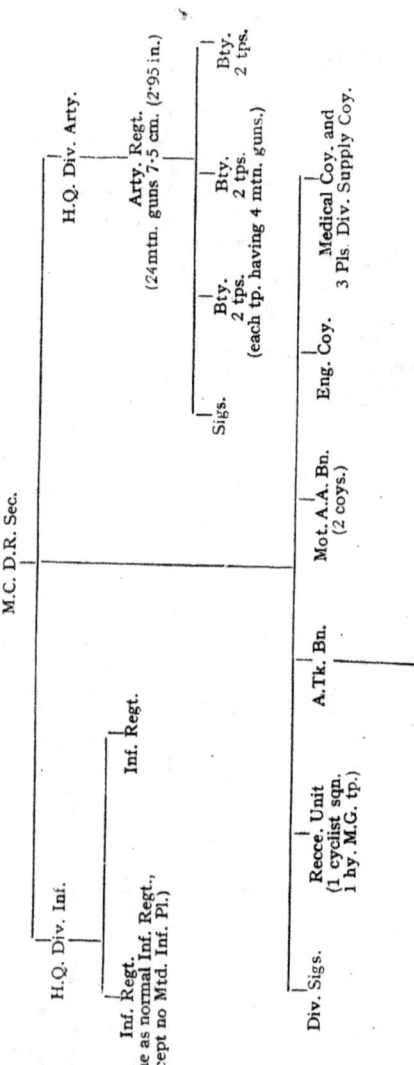

NOTES ON AIR-LANDING TPS.

1. Air-Landing Tps. are normal Inf. carried by plane. The only special training which they receive is practice in emplaning and deplaning. Unlike parachutists, they are not members of the German Air Force.

2. The organization of the Div. may vary, but the following are the main features in which its make-up is likely to differ from that of a normal Inf. Div.

(a) 40 per cent. of strength.

(b) Two instead of three Inf. Regts.

(c) Higher proportion of riflemen and lower of supporting weapons.

(d) Little arty. heavier than 7·5 cm. (2·95 in.) mtn. guns, but it is now possible to transport some heavier weapons including 5 cm. (1·97 in.) A.Tk. guns in small numbers.

(e) Higher proportion of offrs. and N.C.Os.

(f) Skeleton supply services.

3. It is thought that 1,000 tps. could be landed under favourable conditions in 1 hr. on one aerodrome.

4. Air-landing troops will normally be preceded by parachute and glider-borne troops.

30

PARACHUTE REGT.
(Air Force Tps.)
(Strength: approx. 2,200 men)

```
                                    H.Q.
     ┌───────────────┬───────────────┼───────────────┬───────────────┐
   Sigs.            Bn.             Bn.             Bn.
                                (approx. 600 men)
                 ┌────────┬────────┬────────┬────────┐
               Sigs.   Rifle    Rifle    Rifle    Hy. Coy.
                       Coy.     Coy.     Coy.
                   (approx. 144 men)
              ┌─────┬─────┬─────┬─────┐        ┌─────┬─────┬─────┐
            Sigs.  Pl.   Pl.   Pl.              Sigs. M.G. Pl. M.G. Pl.
     ┌────────┬────────┬────────┬────────┐   ┌──────┴──────┐
  Lt. Mortar A.Tk.   L.M.G.  L.M.G.  L.M.G.  M.G. Sec.   M.G. Sec.
    Sec.    Rifle    Sec.    Sec.    Sec.   (2 M.Gs. on
            Sec.   (2 L.M.Gs.)              hy. mountings)
```

13 (Mtn. Gun) Coy.

14 (A.Tk.) Coy.
(incl. Pionęer Pl.)

M.G. Pl.
┌──────┴──────┐
Hy. Mortar   Hy. Mortar Pl.
Sec.   ┌──────┴──────┐
(2 hy. mortars) Hy. Mortar Hy. Mortar
                Sec.   Sec.

NOTES ON PARACHUTE TPS.

1. Besides three Parachute Rifle Regiments, Germany possesses a Parachute Arty. Bty., A.Tk. Bn., M.G. Bn., A.A.M.G. Bn., and Engineer Bn.

2. One Coy. (strength 144) is carried in a Sqn. of 12 Aircraft and trained to land in an area ¼ mile long by 170 yards wide and rally in 12-15 mins. They jump from 300 ft., and during the first few minutes after landing have no longer range weapon than a tommy-gun; the remaining weapons and equipment are landed by parachute in containers distinguished by unit markings. The efficient model 41 A.Tk. gun is a standard weapon in parachute formations.

3. Major demolitions such as destruction of tunnels or steel girder bridges cannot be carried out because of lack of compressor equipment. A.Tk. mines in small numbers are taken.

4. The maximum number of parachute and glider-borne troops which the Germans could use in a single operation is estimated at 12,000. There is, however, no doubt that a very much larger number of parachute troops has by now been trained.

ORGANIZATION OF A SMOKE REGIMENT

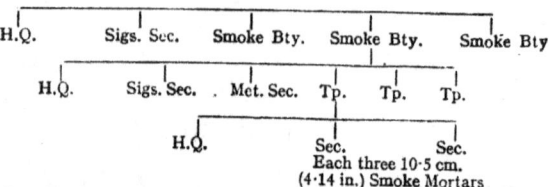

The above is the normal organization, giving 54 smoke mortars to the regiment. Troops of 8 mortars are also known, and one regiment in Russia had some troops of 6 and some of 8: the regimental total might thus rise to anything up to 72 mortars. The peace strength of a battery was 565 all ranks; of a troop, 172 all ranks.

Independent batteries also occur; these may be allotted singly to corps, or a number may be grouped together under a special regimental staff (*Regimentsstab der Nebeltruppe*). Regiments equipped with the 15 cm. (5·91 in.) smoke mortar "d" have the organization given above.

ORGANIZATION OF A DECONTAMINATION (CONTAMINATION) BATTERY

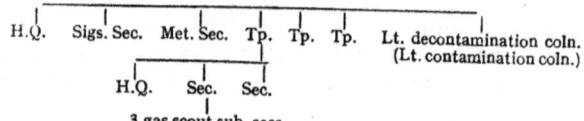

3 gas scout sub. secs.
6 medium decontamination vehicles.
(3 reconnaissance sections
6 medium bulk contamination vehicles)

The decontamination battery can be converted at short notice to the contamination battery by exchanging its decontamination vehicles for bulk contamination vehicles; until needed, these travel in the light decontamination equipment column (2nd line transport, not an integral part of the bty.).

Batteries may be allotted singly to corps, or may be grouped together under a special regimental staff (*Regimentsstab der Nebeltruppe z.b.V*).

Details of the road decontamination battery (*Strassenentgiftungsabteilung*) are not available, but its organization is believed to be much the same as that given above.

Troop decontamination coys. (*Truppenentgiftungskompanien*) are not part of the smoke troops, but are medical troops. Details of their organization are not available.

HINTS ON RECOGNITION OF UNITS AND APPRECIATION OF TACTICS

1. Mounted Troops
   - (a) Mounted or horse-drawn tps.—almost certainly Inf. Div.
   - (b) Horsed Arty. Offrs. and Inf. Offrs. down to Coy. Comds. are, or may be, mounted.
   - (c) Mounted Inf. Pl. of Inf. Regt.—R.H.Q. probably not far away; may be escort for Regtl. Comd.
   - (d) Mounted party of more than 32—Inf. Div. Recce. unit. If held up will get quick support from Hy. Sqn.

2. Cyclists
   Inf. Div. Recce. unit. Will fight for information supported by Hy. Sqn. Used at night in preference to mounted Sqn.

3. Motor-cyclists
   - (a) Solo m/cs. unaccompanied by combinations—probably D.Rs.; their documents and information may be valuable.
   - (b) M/C combinations suggest Armd. Div. or Mot. Inf. Div. BUT 20 in Inf. Div. Recce. Unit.

4. Armd. Cars
   - (a) Total in Inf. Div.—3 Lt. Armd. Cars in Recce. Unit.
   - (b) In Armd. and Mot. Div. Recce. Units the Armd. Car Sqns. each have 1 heavy tp. of 6 Armd. Cars, 1 lt. tp. of 6 Armd. Cars, and 1 lt, tp. of 8 Armd. Cars.

5. Arty.
   - (a) Horse-drawn Arty. must belong to an Inf. Div,

6. Tanks. (*See* Table for details)	(*a*) **Lt. Tanks** in H.Q. Lt. Tank Pls., generally for local recce. H.Q. may be near. (*b*) **Med. Tanks** operate in tps. of 4 ; move fast and support each other. Do not normally fire on the move. Will not usually be committed until A.-Tk. guns have been induced to disclose positions and can be attacked from hull down positions.
General	If Tanks are held up, dive-bombers, or infantry with their close support weapons and M.Gs. will try to restore momentum—during this phase keep your eyes open. Tanks may try to turn flanks.
7. A.-Tk. Defence	(*a*) Areas unsuitable for Tanks will be reinforced with artificial obstacles and mines by Eng.; a minimum of A.-Tk. guns will be sited in these sectors. (*b*) Areas suitable for Tanks will contain bulk of A.-Tk. guns, which will seldom be sited singly. Will probably be in depth, but where penetration must be avoided at all costs, may all be sited to cover forward limit of main defensive position. Infantry are taught not to withdraw because Tanks penetrate their position, but to hold their fire for supporting Inf.
8. Inf.	(*a*) Attack. Effective recce. followed quickly by attack. Sec. leader has a Tommy-gun and will be near the L.M.G.; both will fire during the attack. Inf. guns—loud report and big flash—may give support. (*b*) Defence. Inf. will have alternative and dummy positions and will seldom fire many bursts from the same place. They will try to deceive you about the position of their main defences by vigorous patrolling and throwing out advanced positions. Their favourite defence is counter attack,

IDENTIFICATION

(*See* also " Guide to the Identification of German Units.")

The soldier's unit may be identified from his shoulder-strap, his paybook, or his identity disc.

1. The distinguishing colour [See back cover]

Each arm or service is distinguished by a colour which appears in the form of piping, etc., on the uniforms of officers and O.Rs.

In the case of the field service uniform the colour is worn only as a soutache or inverted dog's leg in front of the cap (F.S.) and as an edging or piping to the shoulder strap, which normally bears also the number of the regiment or equivalent unit.

It should be noted, however, that all of these distinguishing features may be wanting in the field armies. The soutache may be missing from the cap, which in any case is not usually worn in action, and shoulder-straps of O.Rs. have for some time been made and issued without numbers.

The shoulder-straps of both officers and men are detachable, and their withdrawal before any major operation will have to be reckoned with. So far, in Libya, shoulder-straps are still worn by all ranks, but only the officers' straps bear any number.

2. The paybook

This should clearly establish O.Rs. present unit and company, which will be seen at a glance on page 4, section C. For that reason alone the paybook should be taken from P/W and sent back with the prisoners' escort; in any case, as soon as possible.

Page 4 of the paybook provides spaces for the entry of

- A. The last recruiting office to which the soldier was subordinate.
- B. The soldier's late depot unit.
- C. His present company and unit.
- D. His present depot unit.

The unit stamp will also be seen on pages 2 and 23.

3. The identity disc

This should be worn by all officers and men. It is of metal, divided lengthwise by perforations, each half bearing the same inscription. It should be retained by P/W.

The disc shows the man's number, his unit and sub-unit, and a letter, or two letters, indicating his blood-group.

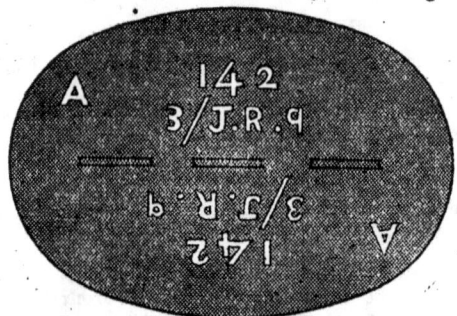

Unfortunately the identity disc is in many cases not up to date, indicating the depot unit where he was trained, but not his actual unit.

4. If all else fails, look for stamps on equipment and clothing.

IF POSSIBLE BASE IDENTIFICATION ON MORE THAN ONE OF ABOVE SOURCES.

5. The following P.W. are wanted quickly :—

        (*a*) Eng.
        (*b*) Recce. unit personnel.
        (*c*) Smoke bn. personnel.
        (*d*) Sigs. (divisional and higher).

6. ALWAYS report details of FLAGS and vehicle markings.

HOW TO REPORT CAPTURE OF PRISONERS

Report immediately, by wireless if possible, giving the following information :—

What ?   Officer, N.C.O. or man.

Which Unit ?   4 Coy. 27 Inf. Regt. ; *exactly how arrived at* (white piping on shoulder-strap ; paybook 4/J.R. 27).

Where captured ?

When captured ?

Whether evacuated and where to ?

ALWAYS send above information (incl. NAME of P.W.) and DOCUMENTS with escort.

For security reasons, the above should normally be forwarded by wireless only if code or cipher is used. If the information is however so urgent that it would be too late if sent by other means, it may be forwarded by wireless in clear.

A.TK. WEAPONS.

	Weapon	Traverse	Most effective range	Rate of fire practical	Penetration	To whom issued (inch type of tks.)
1	A.Tk. Rifle 39 7·9 mm.	—	Under 300 yards	6–8 r.p.m.	30 mm. at 100 yards at normal.	Rifle Coys.
2	A.Tk. Rifle (Polish) 7·9 mm.	—	Under 300 yards	? 6–8 r.p.m.	27·2 mm. at 200 yards at normal.	? some Rifle Coys.
3	A.A./A.Tk. M.G. 2 cm.	360° on cross-mounting 10° on wheels.	300 yards and under	120 r.p.m.	40 mm. at 100 yards at normal. 18 mm. at 200 yards at 20°.	Armoured Cars. Pz. Kw. II. A.A. M.G. Bn. A.Tk. Bn. A.Tk. Coy. of Inf. Regt. Arty. Regt.
4	A.Tk. gun model 41, 2 cm.	90°	Under 300 yards	8–10	76·2 mm at 100 yards at normal. 60 mm. at 200 yards at 30°.	—
5	A.Tk. gun 3·7 cm.	60°	160 yards	8–10 r.p.m.	37 mm. at 600 yards at normal. 25 mm. at 700 yards at 30°.	Old Pz. Kw. III and T.N.H.P. A.Tk. Coy. of Inf. Regt. A.Tk. Bn. (see gun 4·7 cm.). Hy. Sqn. of Recce. Units. Hy. Coy. of Lorried Inf. Bde.
6	A.Tk. gun S.P.M. 38 (Skoda) 4·7 cm.	30°	400 yards and under	8–10 r.p.m.	60 mm. at 220 yards at 30°.	A.Tk. Bn. (gradually replacing gun 3·7 cm.).
7	A.Tk. gun Q.F. 5 cm.	60°	880 yards and under	7–10 r.p.m.	60 mm. at 250 yds. at 30°. 60 mm. at 1,300 yds. at normal	A.Tk. Bn.
8	5 cm. Tank gun	360°	500 yards and under	?	63 mm. at 200 yards at 30°.	Pz. Kw. III.
9	Tk. gun 7·5 cm.	360°	1,500–3,000 yards H.E. 500 yards A.P.	?	55 mm. at 400 yards at 30°.	Pz. Kw. IV and 7·5 cm. S.P. mtg.
10	A.A./A. Tk. gun 8·8 cm.	360°	2,000 yards and under	15–20 r.p.m.	100 mm. at 400 yards at 30°.	Hy. A.Tk. Bns.

PART II.—RECOGNITION AND DATA OF A.F.Vs.

INTRODUCTORY NOTE

It is imperative that every member of an armoured formation, every man firing an anti-tank weapon, and indeed every fighting soldier, should, without any shadow of doubt, be able to distinguish between enemy and British armoured fighting vehicles. The solution of this problem appears at first sight to be in a thorough knowledge of the design and identification marks of our own tanks. The existence, however, of certain features of design common to tanks of all nations complicates recognition; and the possibility that the enemy may deliberately copy any identification marks that we may devise, makes this solution hazardous and unreliable.

By far the most satisfactory way to overcome the doubt which may exist in the minds of those whose task it is to destroy armoured fighting vehicles would then appear to lie in conversancy with external markings and signals and a general knowledge of the peculiarities in the design of armoured fighting vehicles.

To many individuals who have not had the opportunity to learn otherwise one tank appears much the same as another. There are, however, certain clues to the nationality of various A.F.Vs. National design follows a trend and has its own characteristics and it is just as easy, if not easier, for a man sufficiently keen to learn to recognize immediately an A.F.V. as to tell the different types of motor car.

Instruction should begin with a few simple details of the salient features of British tank design. These points, which are set out and illustrated in Section 1 of this book, should be thoroughly understood before turning to the second phase of instruction which deals with points of enemy and foreign vehicles which are characteristic and form a useful guide for recognition purposes.

The third and fourth sections illustrate by drawings the various types of tanks which may be employed by ourselves and the enemy. When these individual illustrations are being studied each point which has been brought out in the two preceding sections should be carefully noted.

In the Middle East, under desert conditions, it was found that owing to dust and haze the lower portions of the tank were often quite invisible. Again, the comparative lack of features and vegetation enabled A.F.Vs. to be seen at long ranges. Many recognition trials were carried out: that giving the most satisfactory result was by the use of pennants. Two pennants are flown on the W/T mast on every A.F.V. (If there is no W/T set in the vehicle a dummy aerial is fitted to carry the pennants.) These two pennants are flown in different positions according to the "Signal of the Day," e.g. two at the top of the mast; one top and one middle, etc.

The colours of the pennants have no bearing on recognition in general as regards British as opposed to enemy A.F.Vs. They are, however, of use within the regiments purely from a domestic aspect.

The fifth section deals with vulnerable points on A.F.Vs. It is possible to lay down a general policy for picking out vulnerable spots to aim at which apply more or less to any type of A.F.V. These points are described and illustrated in this section.

In conclusion, there is one point which must be remembered. The enemy has in his possession a number of our tanks, which, in the event of an invasion, he might use in limited numbers to confuse our defences. Thus, even when a tank is recognized as friendly, it will be advisable to keep it covered until the recognition signal (whatever that may be) is displayed.

SECTION 1

CHARACTERISTICS OF BRITISH DESIGN

1. No known A.F.V. of foreign design has bogey-wheels of **unequal** size (*see* Fig. 1) as in British Cruiser Tanks Marks I and II and Infantry Tank III.

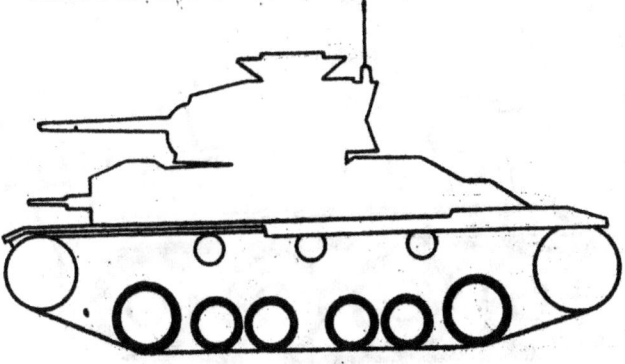

Fig. 1.

2. No tank of enemy design has large Christie-type bogey-wheels (*see* Fig. 2) **unevenly spaced** as in British Cruiser Tanks Marks III, IV and V.

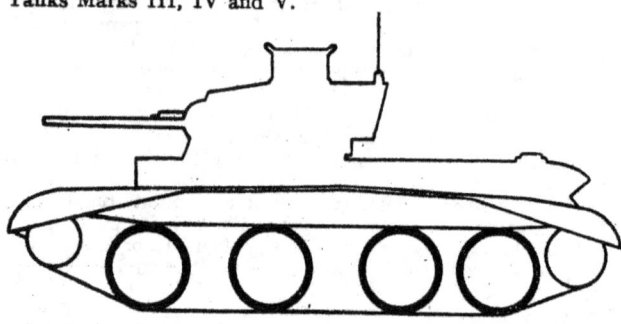

Fig. 2

3. No tanks of enemy design have bogey-wheels which also act as jockey wheels, except some of the early Pz. Kw. II, as in British Light Tanks Mark VII (*see* Fig. 3) and in Cruiser Tanks Marks III, IV, V and VI. Care must be taken, however, not to confuse the Czech L.T.H. (formerly known as T.N.H.P.) Light Tank with the British types. It is being extensively used by the Germans, has four large bogey wheels and is distinguishable by the heavy rivets reaching almost to the top of the tank. However, in the case of L.T.H. there are three jockey wheels, one in between each bogey wheel.

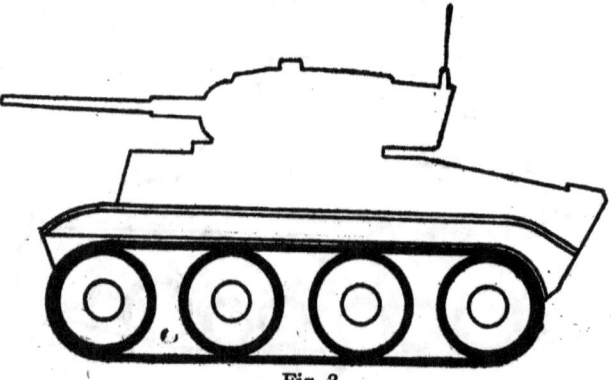

Fig. 3

4. The armour skirting protecting the suspension of British " I " Tank Mark II has five mud chutes (*see* Fig. 4). The skirting seen on some German heavy tanks has only four mud chutes. Similar skirting on French tanks has no mud chutes (*see* Fig. 14).

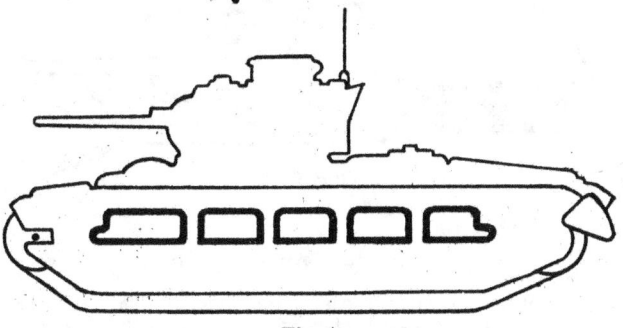

Fig. 4

5. Wireless masts on British tanks are carried on the back of the turret (*see* Fig. 5) which is not the case in foreign designs, where the mast is carried on the hull and away from the turret.

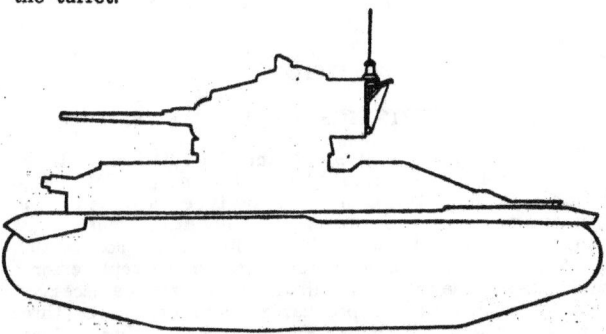

Fig. 5

6. No British tank has outside girder bearers on its suspension system. Both the C.K.D. V.8.H. Medium and Pz. Kw. I Light Tanks, which exist in large numbers in the enemy formations, also the Italian L.33/35, have this distinguishing feature (*see* Fig. 10).

7. The undercut turret (*see* Fig. 6) is a striking peculiarity in some British designs.

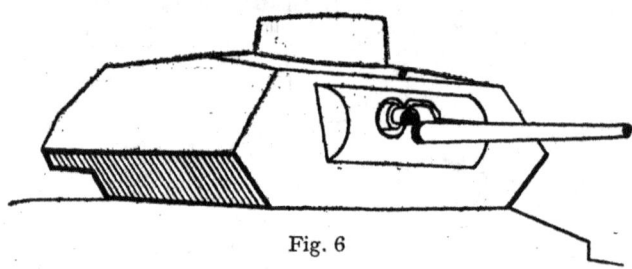

Fig. 6

Armoured cars

The recognition of British Armoured Cars presents very little difficulty.

All British armoured cars have four wheels with the exception of the Lanchester, which is now obsolete and unlikely to be encountered.

All enemy armoured cars have six or eight wheels with the exception of one German car, which has four wheels, but it is open and has no turret, with the exception of one model. This four-wheeled car will be seen in large numbers.

Some British and foreign armoured cars are illustrated in Section 3.

SECTION 2

CHARACTERISTICS OF ENEMY TANKS

There are approximately twenty-six marks of tanks available for the German army. This surprisingly large number includes, of course, many types acquired from German occupied territories. It is doubtful whether the German army will use vehicles other than those which exist in sufficiently large numbers to enable replacements and spares to be readily acquired. It is therefore necessary only to deal with the peculiarities and designs of three different origins—German, French and Czechoslovakian. Their general characteristics are set out in this section, and can be compared with more detailed drawings in Section 4.

i. Tanks of German origin

(a) German designed tanks are low and squat in appearance, not to be confused with some later types of British tanks which may be distinguished by an undercut turret, *see* Fig. 6, Section 1.

(b) Turrets are streamlined and rounded in front to form a roller type gun mantlet. (*See* Fig. 7.) Later models have the roller of the gun mantlet protected by a curved armoured plate fitting closely to the front plate of the turret.

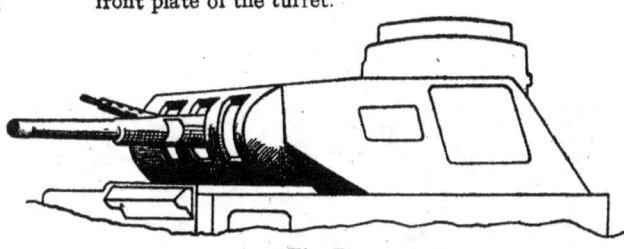

Fig. 7

(c) All medium tanks have a commander's look-out cupola situated at the extreme rear of the turret (*see* Fig. 8). This cupola is circular and is built into the back of the turret, appearing rather as though it was an afterthought.

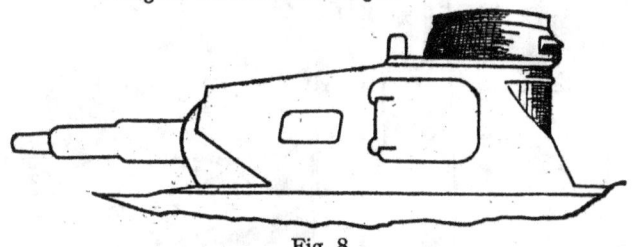

Fig. 8

(d) The back of German tanks where the power unit is housed is flat and usually box-shaped, not protruding very much above the level of the running board.
(e) Their armour plates are mostly welded and the absence of rivets is a feature which is very noticeable at close range and can often be seen through binoculars.

ii. Tanks of Czechoslovakian origin

(a) Heavy riveting of armoured plates is a distinguishing feature which can be seen from quite a long distance (*see* Fig. 9).
(b) The driver's position is on the right or left of the hull with a L.M.G. in a ball mounting. The mounting of the L.M.Gs. are always in ball type mountings and are surrounded by a circular plate heavily riveted (*see* Fig. 9).

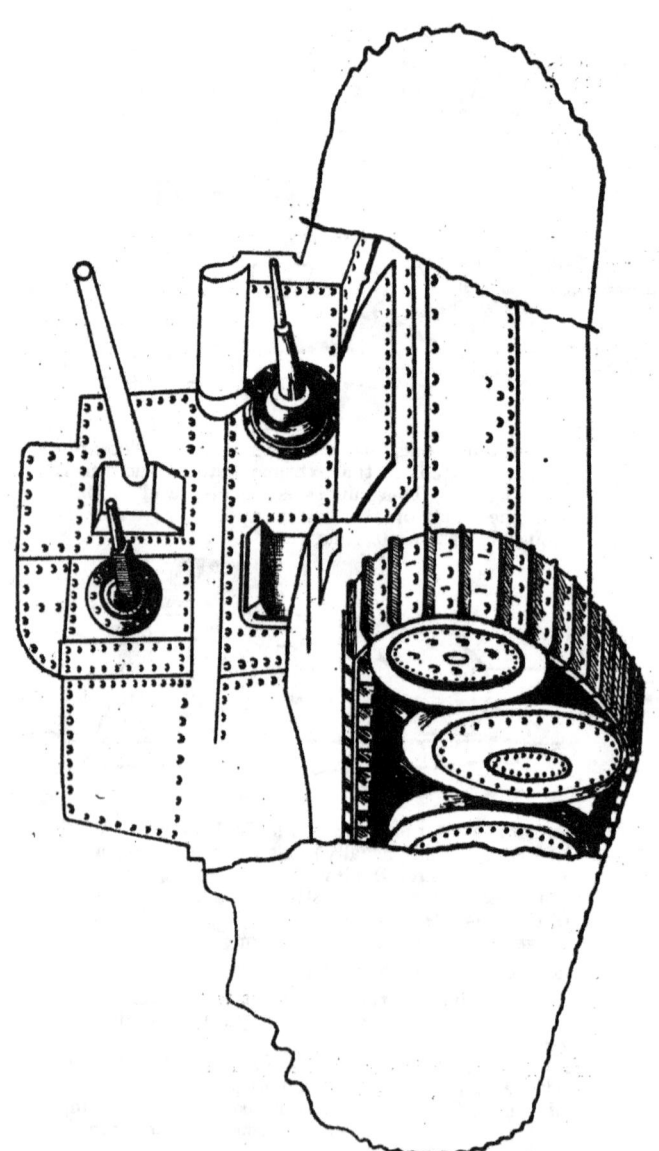

Fig. 9

(c) The power units are situated at the rear and the compartment is built up considerably higher than in German or British tanks and slopes away in the manner of a shallow roofing (see Fig. 11).

(d) Some medium tanks have heavy outside girder bearers on the suspension (see Fig. 10).

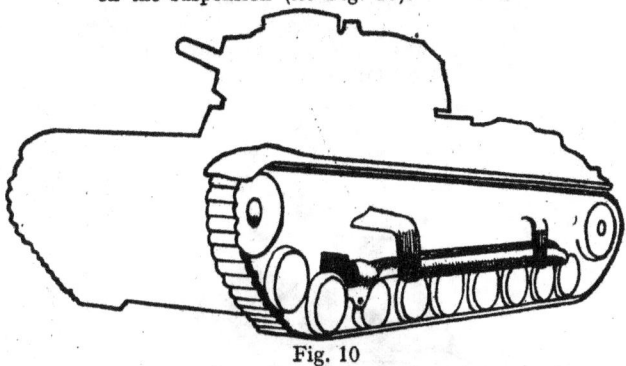

Fig. 10

(e) The idler and sprocket wheels often have a large flange which protrudes over the edge of the tracks. The object of the flange is to prevent the track slipping off, and more especially to keep the track pins, which are of the floating type, in place (see Fig. 11).

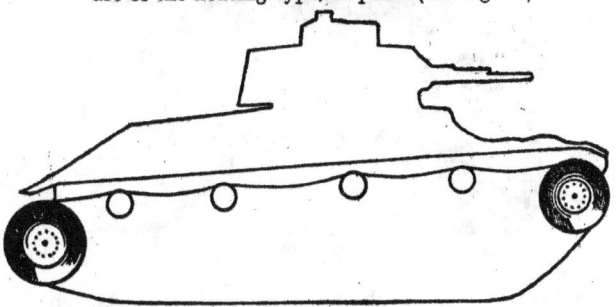

Fig. 11

iii. Tanks of French origin

(a) French tanks are particularly noticeable because most of their armour plate is made in huge castings which has the effect of rounding off all edges and angles. There are very few sharp lines or angles on French A.F.Vs. (*see* Fig. 12).

(b) Seen from the front, French tanks present a pyramid form of silhouette. The main casing of the hull is built well up above the top of the tracks and slopes inwards. The turret continues this inward slope, the top of the turret being always narrower than the base. These peculiarities together with the rounded plates give the French tanks the appearance of being very heavily armoured (*see* Fig. 12).

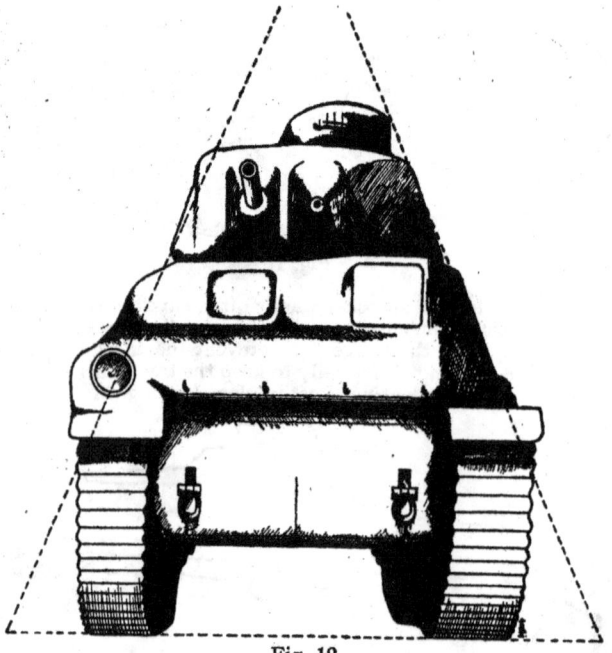

Fig. 12

(c) On tanks where there is a look-out cupola, the cupola is in all cases dome-shaped (*see* Fig. 12). French tanks used by the Germans may have a cupola fitted with two flaps on a flat top.

(d) The engine compartment is even higher than the Czech tanks, but slopes away in a similar fashion.

(e) Scissor type vertical articulation is used in some French A.F.V. suspensions (*see* Fig. 13).

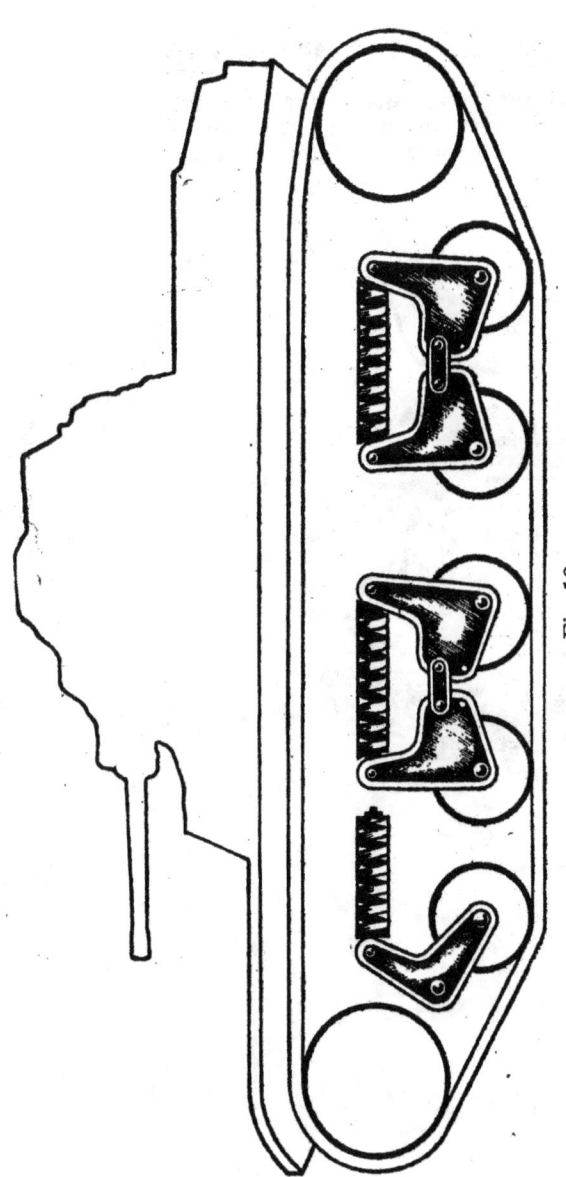

Fig. 13

(f) Medium heavy tank " Char " B has a high heavy track, with suspension protected by armoured skirting. There are no mud chutes in this protective skirting (see Fig. 14).

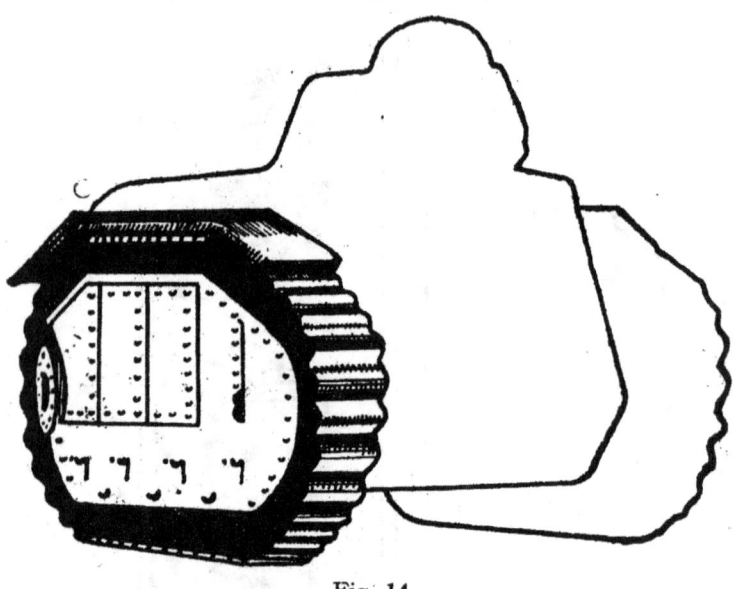

Fig. 14

SECTION 3—DRAWINGS AND SALIENT FEATURES OF BRITISH A.F.Vs.

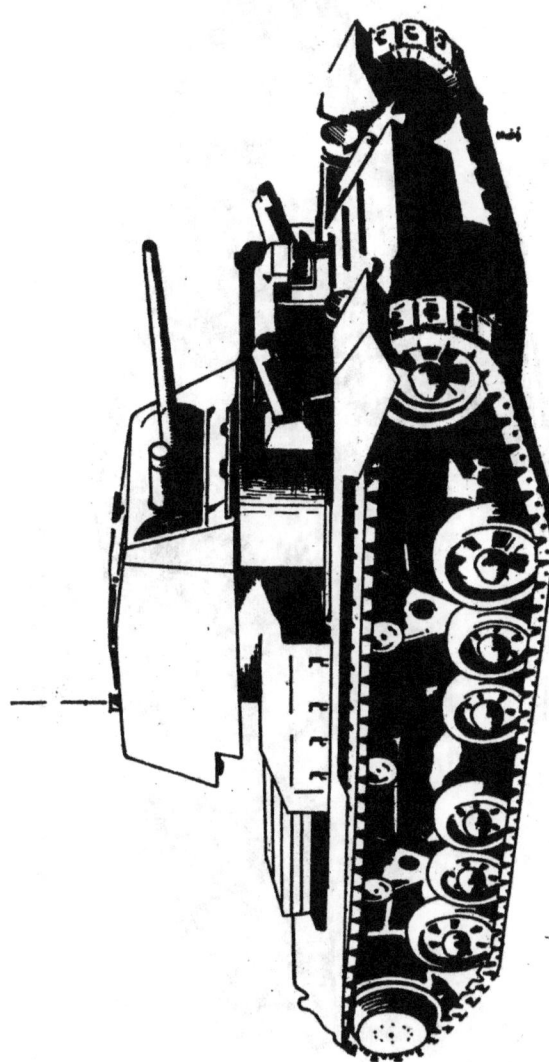

British Cruiser Tank Mark I.

Bogey of unequal size; Square turret; Two round auxiliary M.G. turrets forward; Medium fast. Becoming obsolete.

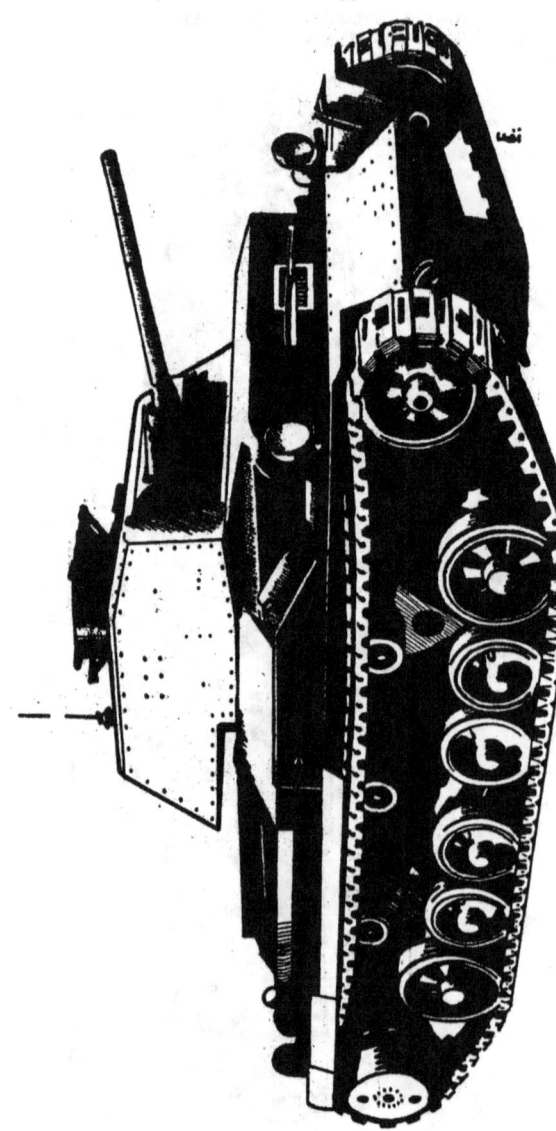

British Cruiser Tank Mark II.
Bogey wheels of unequal size; Square turret; Medium fast. Becoming obsolete.

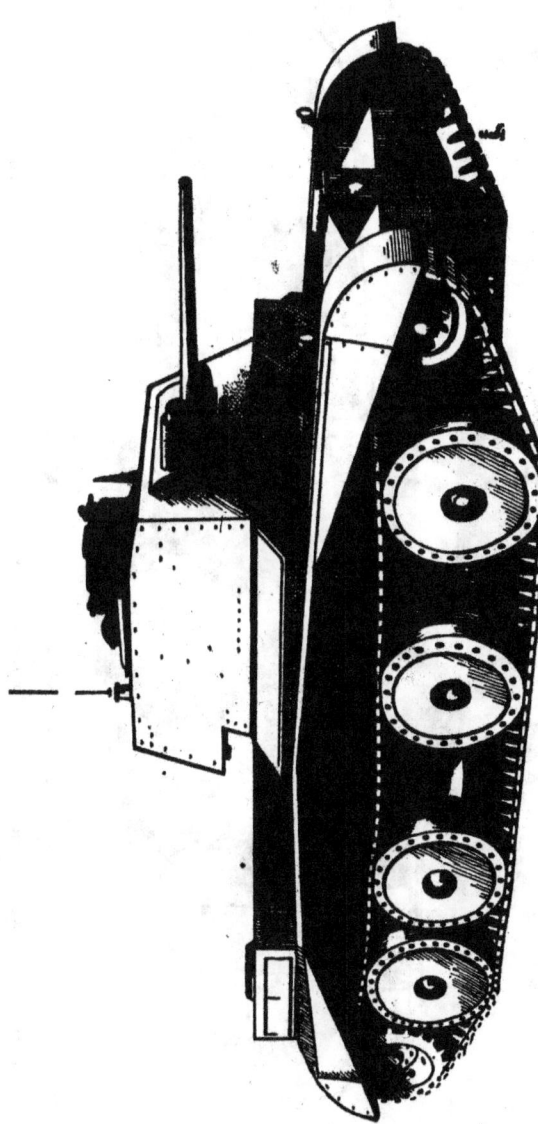

British Cruiser Tank Mark III.

Four large bogey wheels unevenly spaced touching top and bottom of tracks; Square turret; Fast. Obsolete.

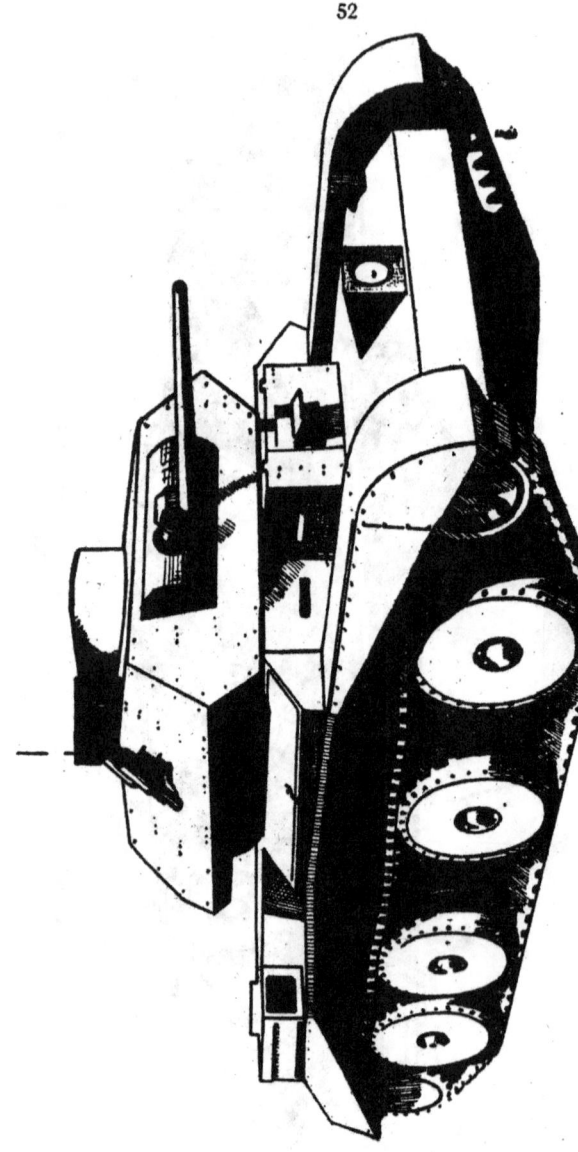

British Cruiser Tank Mark IV.

Four large bogey wheels unevenly spaced touching top and bottom of track; Diamond-shaped turret; Fast.

British Cruiser Tank Mark V. COVENANTER.
Four large bogey wheels unevenly spaced touching top and bottom of track; Square diamond-shaped turret; Twin aerials on back of turret; Fast.

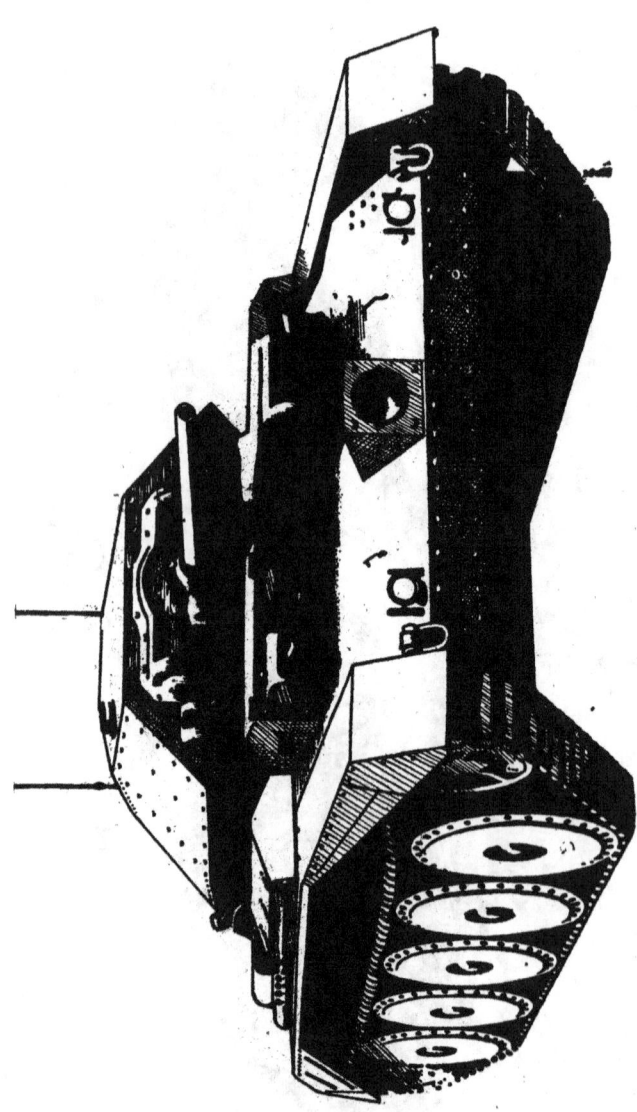

British Cruiser Tank Mark VI. CRUSADER.
Five large bogey wheels touching top and bottom of track; Squat diamond-shaped turret; Twin aerials on back of turret; Fast.

British Infantry Tank Mark I.

Small and low; Armament: 1 Vickers M.G.; Very slow. Obsolete.

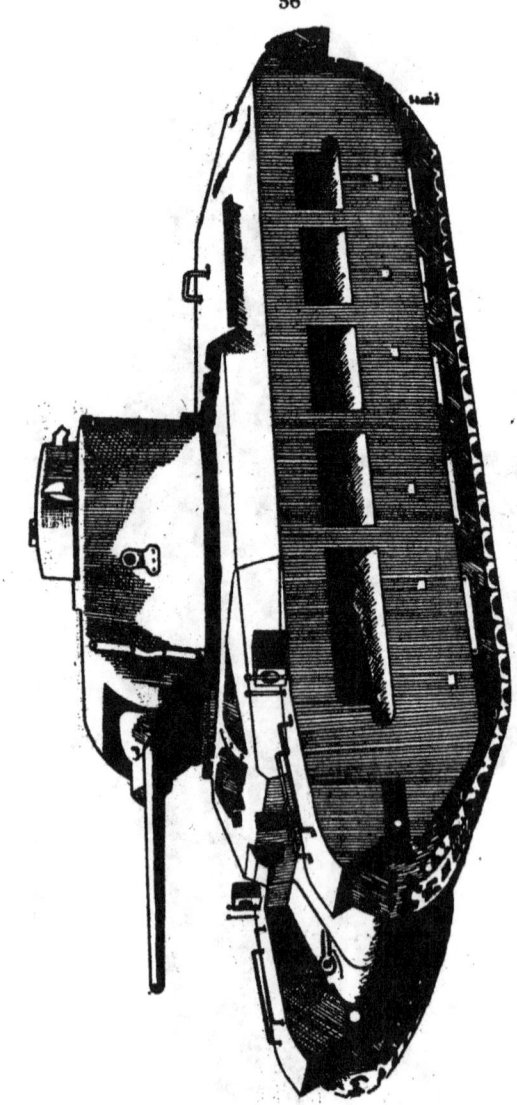

British Infantry Tank Mark II. MATILDA.

Armour skirting covering suspension; Five mud chutes; Distinctive shape of turret; Slow.

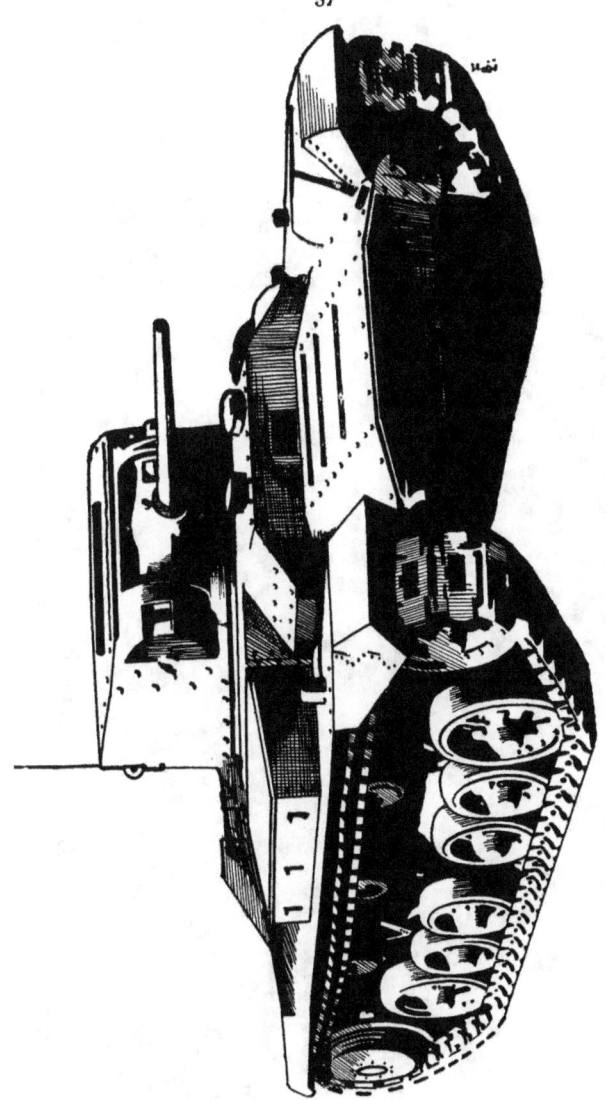

British Infantry Tank Mark III. VALENTINE.
Uneven size of bogey wheels; Rounded shape of turret; Slow.

British Infantry Tank Mark IV. CHURCHILL.
Small wheels with coil spring to each, track over hull, cast turret.

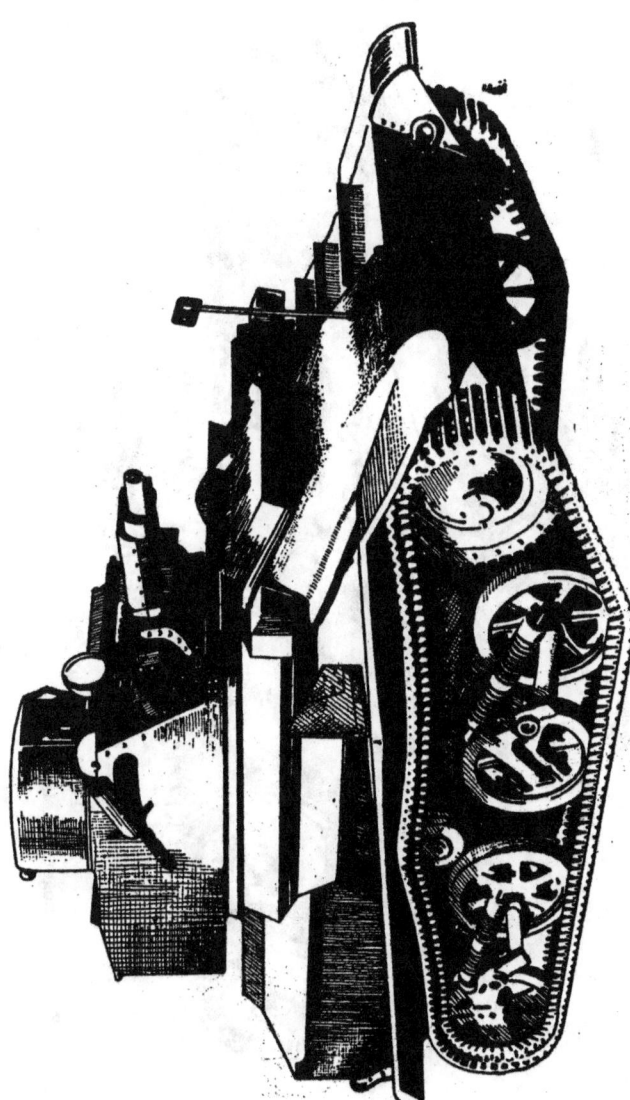

British Light Tank Mark VI (a)
Two pairs of bogies of even size; Fast.

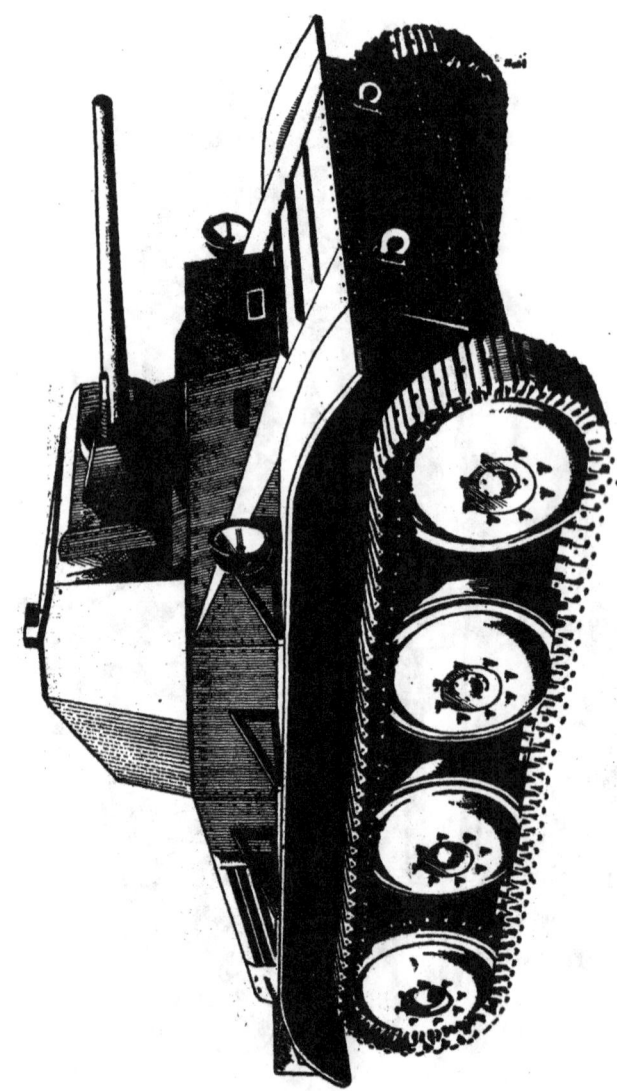

British Light Tank Mark VII. TETRARCH.

Four large bogey wheels touching top and bottom of track; No spocket or idler wheels. <u>Fast</u>.

British Armoured Car Guy Mark I.

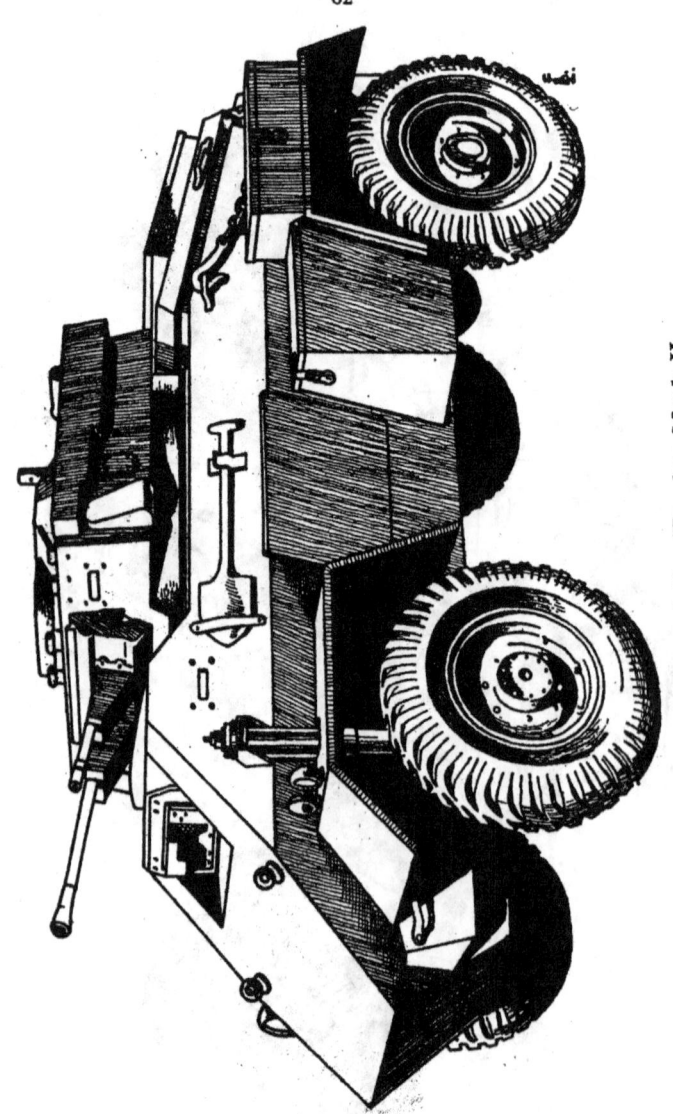

British Armoured Car Humber Mark II.

British Armoured Car Daimler Mark I.

American Light Tank, M.3., GENERAL STUART.
Two pairs of bogies; large rear w... ...l; note rand shields fitted

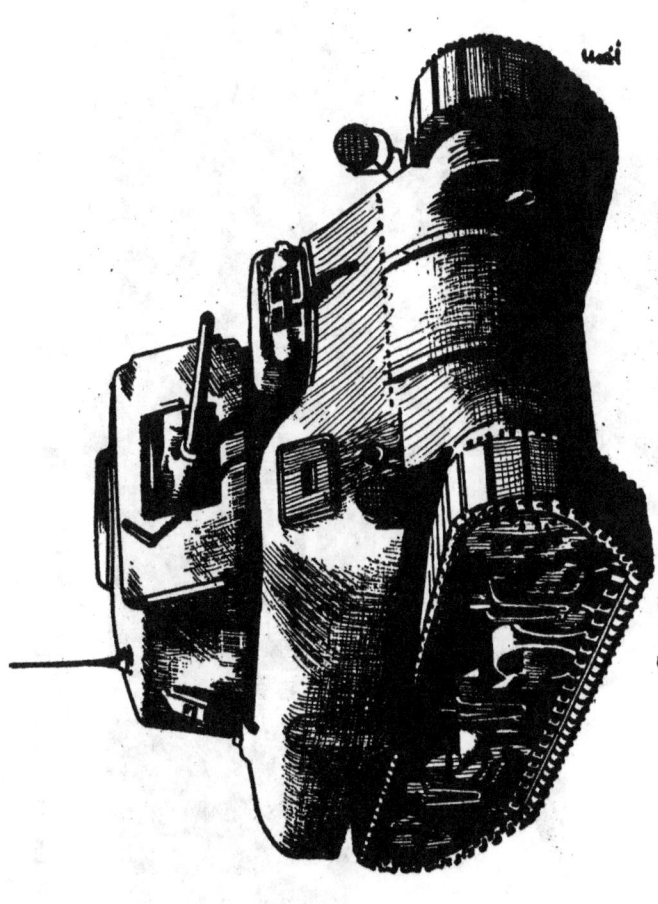

Canadian Medium Tank M.3. **RAM.**

Six bogey wheels in pairs; Rounded plates; Superstructure overlapping tracks; Subsidiary M.G. turret; Ribbed front plates.

American Medium Tank M.3. GENERAL LEE.
Six bogey wheels in pairs; High built up superstructure; 75 mm. gun in subsidiary hull turret; M.G. in look-out cupola; Ribbed front plates.

American Medium Tank M.3. GENERAL GRANT.

SECTION 4

DRAWINGS AND SPECIFICATIONS OF ENEMY A.F.Vs.

German Pz. Kw. I Light Tank

Weight:	5·7 tons.
Crew:	2 men.
Armour:	Front 18 mm. Turret 18 mm. Remainder 14 mm.
Armament:	2–7·91 mm. M.Gs. coaxially mounted.
Ammunition carried:	2,000 rounds.
Dimensions:	Length, 12 ft. 6 in. Width, 8 ft. 0 in. Height, 5 ft. 7 in. Ground clearance, 12 in.
Performance:	Trench, 4 ft. 7 in. Step, 1 ft. 2 in. Water, 2 ft. 0 in. Gradient, 45°. } Estimated.
Engine:	90/100 h.p
Speed:	Road, 32 m.p.h; radius of action 95 miles.
Fuel:	Petrol.
Suspension:	5 bogey wheels. Combination of rocker and semi-elliptical springs and independent coil springs.
Communication:	External, W/T.

German Pz. Kw. II Light Tank

Weight:	9 tons.
Crew:	3.
Armour basis:	Front, 35 mm. gun mounting and front of turret. Superstructure, front, 35 mm.; remainder, 15–18 mm.
Armament:	1 2 cm. H.M.G.; 1 L.M.G. coaxially mounted.
Ammunition carried:	224 rounds, for H.M.G.
Dimensions:	Length, 15 ft. 4 in. Width, 7 ft. 2 in. Height, 6 ft. 5 in. Ground clearance, 0 ft. 11 in.
Performance:	Trench, 4 ft. 11 in. Step, 1 ft. 11 in. Water, 2 ft. 6 in. Gradient, 45°. } estimated.
Engine:	Maybach, 6 cyl.
Speed:	Road, 25 m.p.h.; radius of action 125 miles.
Fuel:	Petrol.
Suspension:	5 independent elliptically sprung bogey wheels.
Observation:	8 periscopes for gunner in cupola.
Communication:	Internal, telephone. External, W/T.

German Pz. Kw. III Light Medium Tank

Weight:	18–20 tons.
Crew:	5: driver, hull gunner/W/T. operator, commander, gunner and loader.
Armour basis:	Gun mounting, 70 mm.; front superstructure, 60 mm.; rear deck, 20 mm. remainder, 30 mm. There is also a model with all front armour 50 mm.
Armament:	1 5 cm. Q.F. gun; 1 7·91 mm. M.G. coaxially mounted; 1 7·91 M.G. in hull.
Ammunition carried:	100 rds. for Q.F. gun; 2,000 rds. for M.Gs.
Dimensions:	Length, 17 ft. 8 in. Width, 9 ft. 7 in. Height, 7 ft. 9 in. Ground clearance, 1 ft. 0 in.—9 in.
Performance:	Trench, 6 ft. to 7 ft. Step, 3 ft. 0 in. } estimated. Water, 2 ft. 11 in.
Engine:	Maybach V.12 cyl., 320 h.p.
Speed:	28 m.p.h.; radius of action, 75–100 miles (estimated).
Fuel:	103 gallons petrol.
Suspension:	6 independent bogey wheels.
Observation:	All-round periscopes.
Communication:	Internal, telephone. External, W/T.

German Pz. Kw. IV Medium Tank

Weight :	22 tons.
Crew :	5.
Armour basis :	Sides of hull, superstructure 40 mm., front plates 60 mm., rear deck 10 mm., remainder 20 mm.
Armament :	1 7·5 cm gun ; 1 7·91 mm. M.G. coaxially mounted ; 1 7·91 mm. M.G. in hull.
Ammunition carried :	85 rounds H.E. & A.P. for gun, 2,000 rounds for M.Gs.
Dimensions :	Length, 18 ft. 6 in. Width, 9 ft. 3 in. Height, 8 ft. 7 in. Ground clearance, 1 ft. 2 in.
Performance :	Trench, 6 ft. to 7 ft. Step, 3 ft. } estimated. Water, 3 ft. 0 in.
Engine :	Maybach V.12, 320 h.p
Speed :	Road, 23 m.p.h. ; radius of action 75–100 miles (estimated).
Fuel :	103 gallons petrol.
Suspension :	8 small bogey wheels in pairs with cantilever springs.
Observation :	All-round periscopes
Communication :	Internal, telephone. External, W/T.

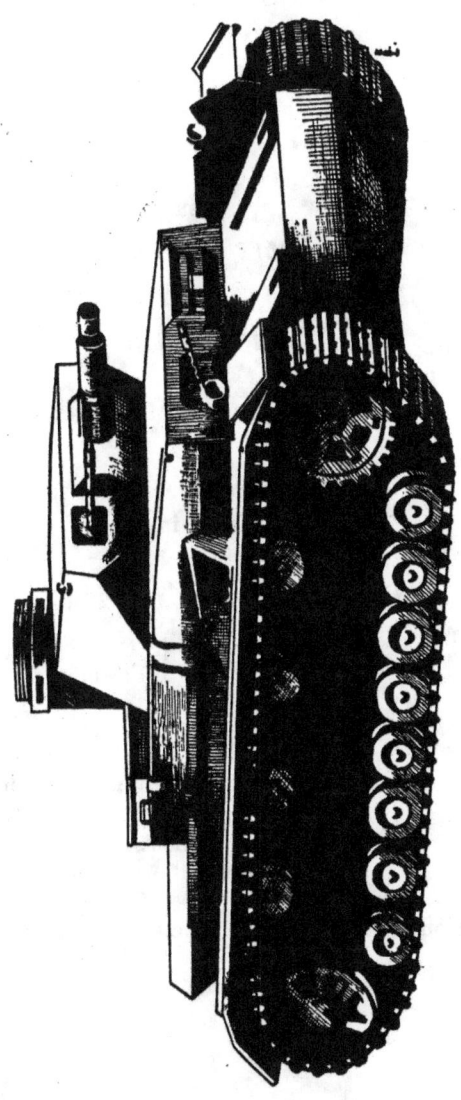

(ex-Czechoslovakia)

Pz. Kw. 35/T (L.T.35)

Weight :	10·3 tons.
Crew :	4.
Armour basis :	25 mm.
Armament :	1 3·7 cm. gun, 1 L.M.G. coax, 1 L.M.G. in hull.
Ammunition carried :	90 rds. for gun, 3,000 rds. for M.Gs.
Dimensions :	Length, 14 ft. 11 in. Width, 7 ft. Height, 7 ft. 3 in. Ground clearance, 1 ft. 2 in.
Performance :	Trench, 6 ft. 6 in. Step, 2 ft. 7 in. Water, 2 ft. 7 in. Gradient, 28° } Estimated.
Engine :	120 h.p. 6 cyl.
Speed	Road 22 m.p.h., radius of action 72 miles.
Suspension :	9 small bogey wheels, 2 pairs of twin bogies on rocker with leaf springing, 1 independent bogey wheel.
Communication :	External W/T receiver and transmitter and lamp.

(Ex-Czechoslovakia)

Pz. Kw. 38/T (L.T.H. Light Tank.)

Weight :	9·5 tons.
Crew :	4.
Armour basis :	25 mm. (now 50 mm.).
Armament :	1 4·7 cm. gun ; 2 M.Gs.
Ammunition carried :	90 rounds for gun ; 3,000 for M.Gs.
Dimensions :	Length, 16 ft. 1 in. Width, 6 ft. 9 in. Height, 7 ft. 9 in. Ground clearance, 1 ft. 4 in.
Performance :	Trench, 6 ft. 5 in. Step, 2 ft. 8 in. Water, 2 ft. 11 in. Gradient, 28°.
Engine :	125 h.p., 6 cyl., water cooled.
Speed :	Road, 35 m.p.h. ; radius of action 125 miles.
Suspension :	4 large bogey wheels, semi-elliptical springing.
Communication :	External, W/T.

German (ex-Czechoslovakia) C.K.D. V 8 H Light medium

Weight:	16·5 tons.
Crew:	4.
Armour basis:	36 mm.
Armament:	1 4·7 cm. gun., 1 M.G. coax., 1 M.G. in hull.
Ammunition carried:	90 rds. for gun, 3,000 for M.Gs.
Dimensions:	Length, 17 ft. 6 in. Width, 7 ft. 6 in. Height, 7 ft. 8 in. Ground clearance, 1 ft. 6 in.
Performance:	Trench, 7 ft. 6 in. Step, 3 ft. 3 in. Water, 3 ft. 3 in. Gradient, 41°.
Engine:	245 h.p. V.8 cyl.
Speed:	Road, 27 m.p.h.; radius of action, 77 miles.
Fuel:	Petrol.
Suspension:	9 bogey wheels, 1 independent, 4 pairs with semi-elliptical leaf springing.
Communication:	Internal, telephone. External, W/T.

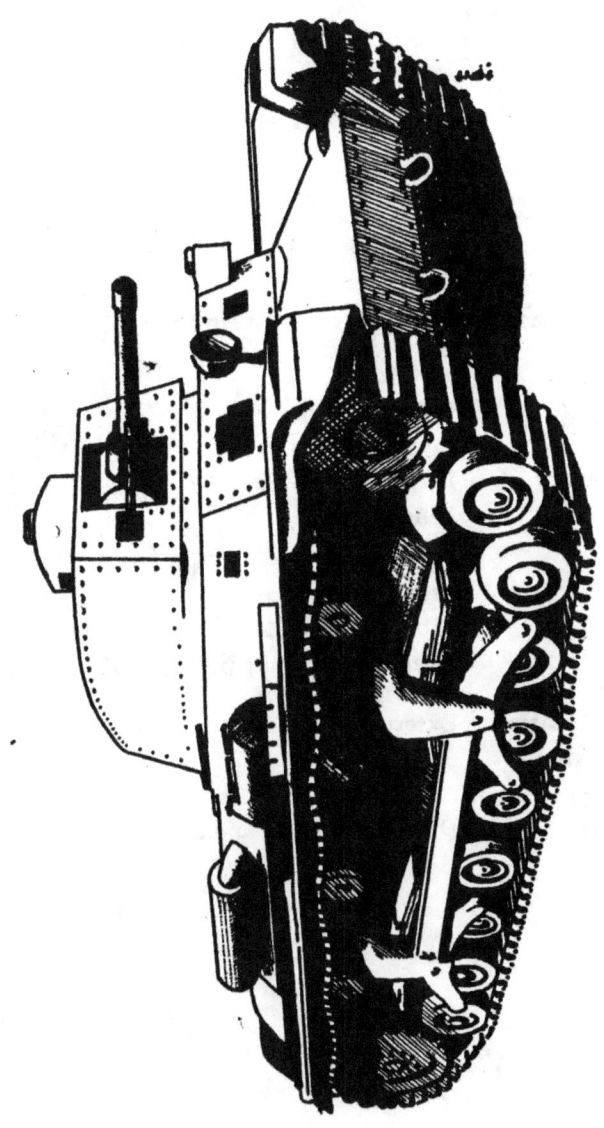

German Heavy 8-wheel Armoured Car

Weight:	9–10 tons.
Crew:	4–5.
Armour basis:	Front 15 mm., sides 8 mm.
Armament:	1 2 cm. H.M.G. or 1 3·7 cm. gun, 1 L.M.G.
Dimensions:	Length, 21 ft. Width, 7 ft. Height, 7 ft. 6 in. Ground clearance, 9 ins.
Engine:	155 h.p. V 8 cyl.
Speed:	Road, 28–31 m.p.h.; radius of action 124–155 miles.
Communication:	External W/T and flag.
No. of wheels:	8.
Drive:	All 8 wheels.
Steering:	Dual control fore and aft.

German Light Armoured Car. Sd. Kfz. 222

Weight:	4·7 tons.
Crew:	3.
Armour basis:	Visors 15 mm., remainder 8 mm.
Armament:	1 2 cm. H.M.G., 1 M.G. Both can be elevated for A.A. defence.
Dimensions:	Length, 15 ft. 7 in. Width, 6 ft. 3½ in. Height, 7 ft. 4½ in. to top of cage. Ground clearance, 7¾ in.
Engine:	75 h.p., 8 cyl.
Fuel:	22 gallons petrol.
Speed:	Road, 28–31 m.p.h.; radius of action, 124–155 miles.
Communication:	External, W/T.
No. of wheels:	4.
Drive:	All four wheels.
Steering:	All four wheels.

German A.S.P.6

Weight:	6·4 tons.
Crew:	4.
Armour basis:	14 mm.
Armament:	1 20 mm. H.M.G., 1 L.M.G.
Dimensions:	Length, 16 ft. 5 in. Width, 7 ft. 6 in. Height, 8 ft.–9 ft.
Engine:	100 h.p., 6 cyl.
Speed:	Road, 28–31 m.p.h.; radius of action, 168–216 miles.
Communication:	External, W/T and flag.
No. of wheels:	6.
Drive:	Rear 4 wheels.
Steering:	Dual control fore and aft.

German leichter Panzer-Spähwagen (Light Reconnaissance Car)

Weight:	1·7 tons.
Crew:	2 (3 when W/T carried).
Armour basis:	10 mm.
Armament:	1 L.M.G. (convertible for A.A. duties).
Dimensions:	Length, 10 ft. Width, 5 ft. 6 in. Height, 5 ft. 3 in. Ground clearance, 1 ft. 6 in.
Engine:	Horch, 6 cyl.
Speed:	Road, 40–50 m.p.h.; radius of action, 175 miles.
Fuel:	Petrol.
Communication:	External, flag, sometimes W/T.
No. of wheels:	4.
Steering:	Front wheels.
Drive:	Rear wheels.

NOTE.—This vehicle is obsolete.

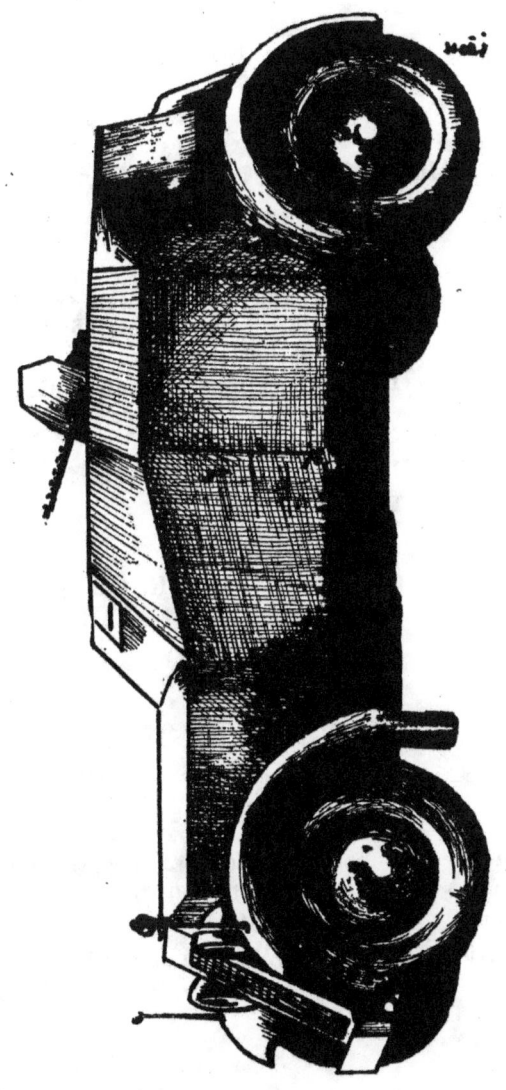

Italian M.13/40 Medium Tank

Weight:	13·5 tons.
Crew:	4.
Armour basis:	Front plate, 30 mm.; turret, sides, 25 mm.; top plates, 15 mm.; rear deck, 8 mm.
Armament:	1 47/32 Q.F. gun, 1-8 mm. M.G. coaxially mounted, 2-8 mm. M.Gs. in hull.
Ammunition carried:	104 rds. for gun, 3,040 rds. for M.Gs
Dimensions:	Length, 16 ft. Width, 7 ft. 3 in. Height, 7 ft. 9 in. Ground clearance, 1 ft. 4 in.
Performance:	Trench, 6 ft. 6 in. Step, 2 ft. 11 in. Water, 3 ft. 3 in. Gradient, 40°.
Engine:	105 h.p. Diesel V.8 cyl.
Speed:	Road, 19 m.p.h.; radius of action, 125 miles.
Fuel:	Diesel.
Suspension:	8 bogey wheels, 2 bogies and semi-elliptic leaf springing.
Observation:	Periscopes, slits.
Communication:	External, W/T, flag or hand.

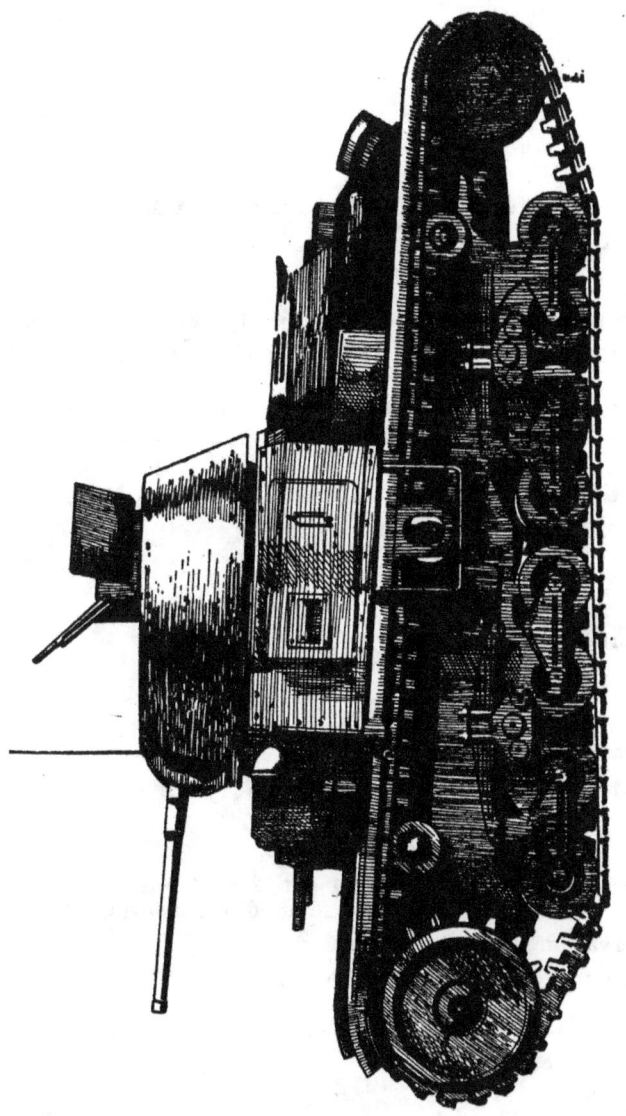

Italian M.11/39 Medium Tank

Weight:	11 tons.
Crew:	3.
Armour basis:	Front and turret, 30 mm.; sides, 15 mm.; belly, 12 mm.; top, 10 mm.
Armament:	1 37/40 Q.F. in hull; 2 8 mm. M.Gs. in turret.
Ammunition carried:	84 rds. for gun, 1,440 rds. for M.Gs.
Dimensions:	Length, 15 ft. 6 in. Width, 7 ft. 2 in. Height, 7 ft. Ground clearance, 1 ft. 2 in.
Performance:	Trench, 6 ft. 7 in. Step, 2 ft. 7 in. Water, 3 ft. 3 in. Gradient, 45°.
Engine:	105 h.p. Diesel.
Speed:	Road, 20 m.p.h.; radius of action, 125 miles.
Fuel:	Diesel, 40 gallons.
Suspension:	2 bogies of 4 bogey wheels each, rocker arm and semi-elliptical springing.
Observation:	Periscopes and slits.
Communication:	External, W/T.

Italian L.33/35 Light Tank

Weight:	3·5 tons.
Crew:	2.
Armour basis:	Front and turret, 16 mm., remainder 4–8 mm.
Armament:	2–8 mm. M.Gs. or 1 H.M.G. and flame-thrower.
Dimensions:	Length, 10 ft. 4 in. Width, 4 ft. 7 in. Height, 4 ft. 3½ in. Ground clearance, 9 in.
Performance:	Trench, 4 ft. 9 in. Step, 2 ft. 1½ in. Water, 2 ft. 2 in. Gradient, 45°.
Engine:	43 h.p. 4 cyl. Fiat.
Speed:	Road, 26 m.p.h.; radius of action, 62 miles.
Fuel:	Petrol.
Suspension:	2 bogies of three wheels each connected by girder bearer. 1 independent.
Observation:	Slits and shutters.
Communication:	External, Flag or W/T.

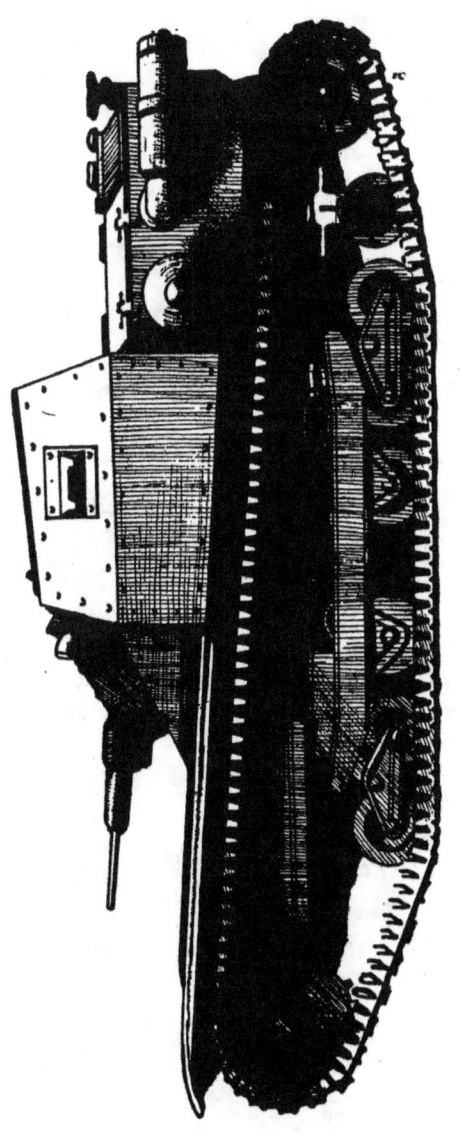

Italian L.6–40 Light Tank

Weight:	6·4 tons.
Crew:	2.
Armament:	1 20 mm. H.M.G. coaxially mounted with 1 8 mm. M.G.
Dimensions:	Length, 12 ft. 5 in. Width, 6 ft. 5 in. Height, 6 ft. 5 in. Ground clearance, 1 ft. 1 in.
Performance:	Trench, 4 ft. 3 in. Step, 2 ft. 11 in. Water, 2 ft. 11½ in. Gradient, 40°.
Engine:	70 h.p.
Speed:	Road, 22 m.p.h.
Fuel:	Petrol.

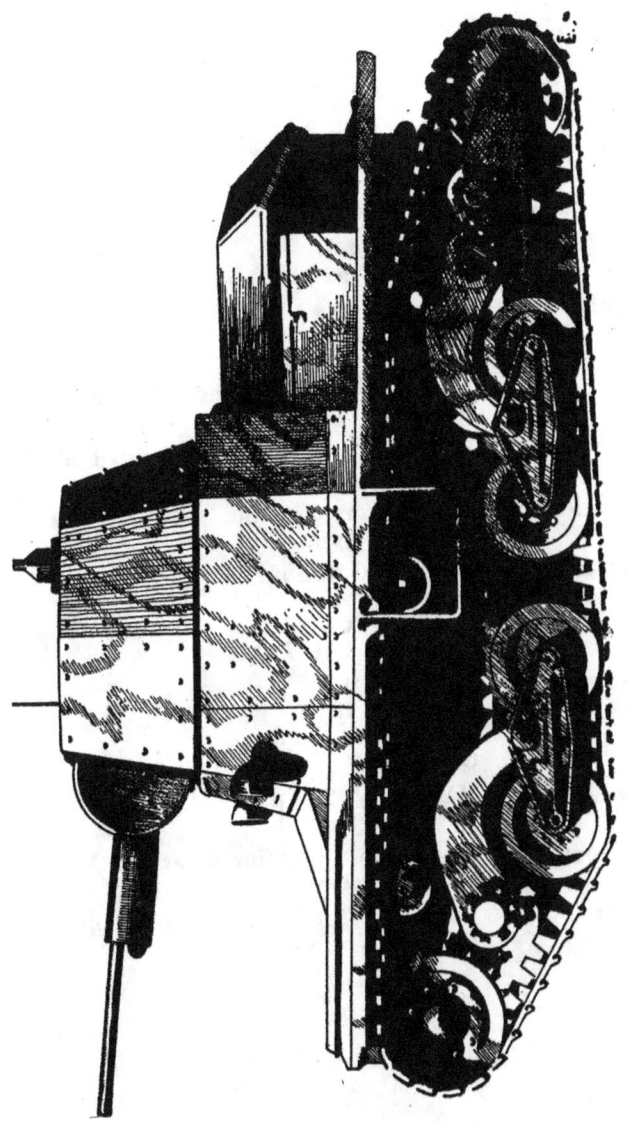

French Char. B I bis.

Weight:	31 tons.
Crew:	4.
Armour:	Front, 60 mm.; sides, 60 mm.; roof, 20 mm.; floor, 20 mm.
Armament:	1 75 mm. gun (mounted in hull); 1 47 mm. gun; 2 7·5 mm. M.Gs.
Ammunition:	80 rds. for 75 mm. gun; 76 rds. for 47 mm. gun.
Dimensions:	Length, 20 ft. 8 in. Width, 8 ft. 2 in. Height, 9 ft. Clearance, 1 ft. 7 in.
Performance:	Trench, 9 ft. Step, 3 ft. 10 in. Water, 4 ft. 10 in. Gradient, 40°.
Speeds:	Maximum, 17 m.p.h. Cruising, 15 m.p.h.
Across country:	11–12 m.p.h.; radius of action, 10 hours or 110 miles.
Control:	External, Wireless, lamps and flags. Internal, Voice tube.

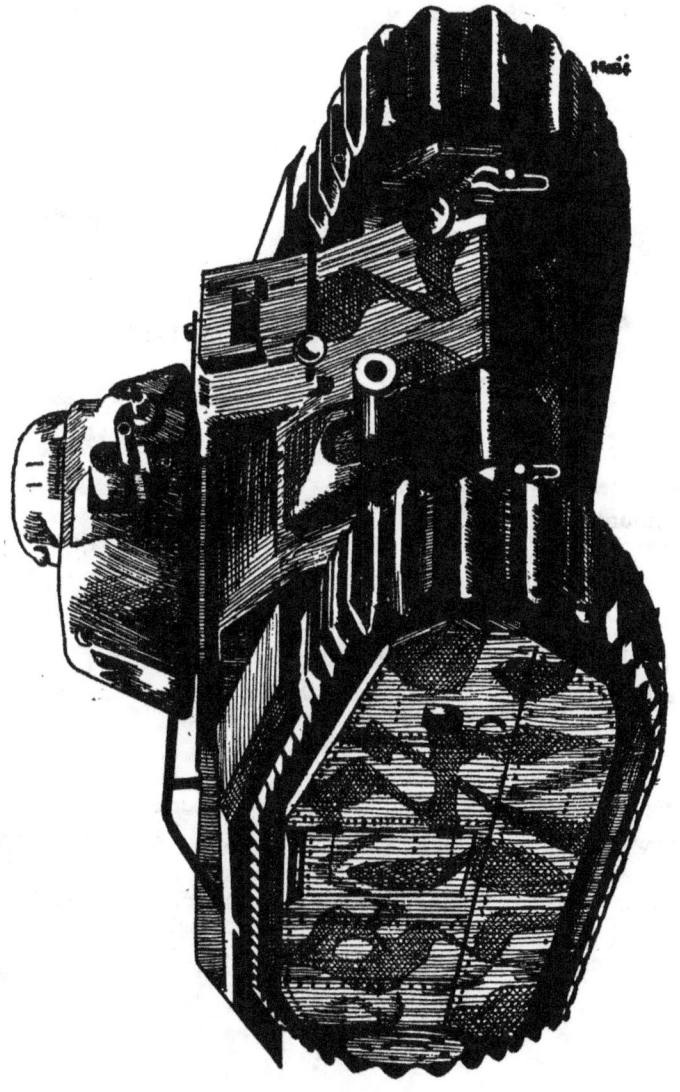

French Somua S.35 Medium Tank

Weight:	18 tons.
Crew:	3.
Armour basis:	40 mm.
Armament:	1·47 mm. gun, 1 L.M.G.
Ammunition carried:	120 rds. for gun; 5,000 rds. for M.G.
Dimensions:	Length, 16 ft. 5 in. Width, 6 ft. 8 in. Height, 8 ft. 10 in. Ground clearance, 1 ft. 4 in.
Performance:	Trench, 7 ft. 10 in. Step, 2 ft. 11 in. Water, 3 ft. 3 in. Gradient, 40°.
Engine:	150 h.p. V.8 cyl. water cooled.
Speed:	Road, 29 m.p.h.; radius of action, 125 miles.
Suspension:	9 bogey wheels, 4 bogies, 1 independent bogey wheel, leaf springing.
Communication:	External, W/T.

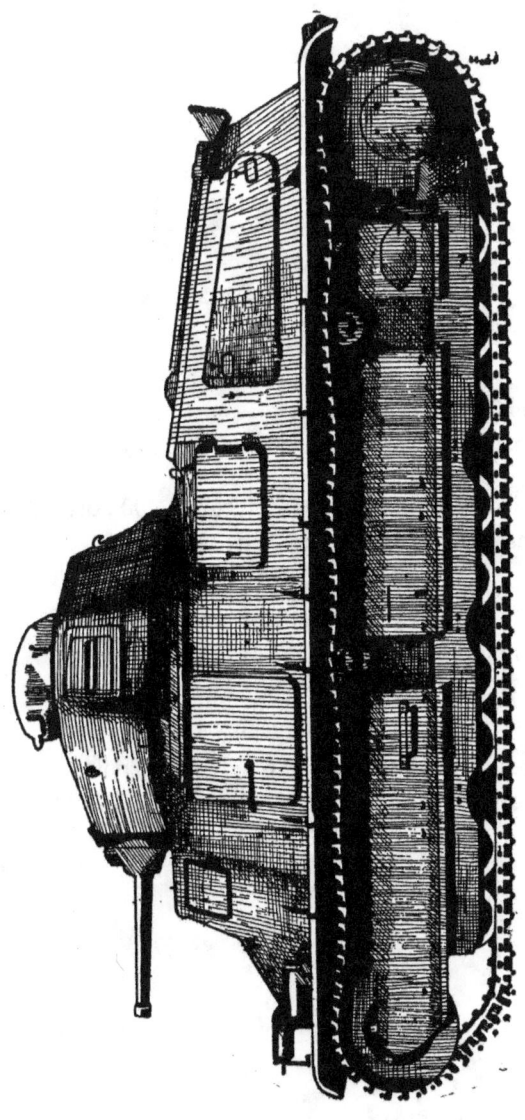

French H.39 Light Tank

Weight.	12 tons.
Crew :	2.
Armour basis :	40 mm.
Armament :	1 3·7 cm. gun, 1 L.M.G. coax.
Ammunition carried :	100 rds. for gun, 2,400 rds. S.A.A.
Dimensions :	Length, 13 ft. 8 in. Width, 6 ft. 1 in. Height, 7 ft. Ground clearance, 1 ft. 3 in.
Performance :	Trench, 4 ft. 11 in. Step, 2 ft. 7 in. Water, 2 ft. 7 in. Gradient, 40°.
Engine :	120 h.p. 6 cyl. watercooled.
Speed :	Road, 26 m.p.h. ; radius of action, 120 miles.
Suspension :	6 bogey wheels ; 3 bogies, scissors articulation.
Communication :	External flag.

French Renault R 35. Light Tank

Weight:	11 tons.
Crew:	2.
Armour basis:	40 mm.
Armament:	1 37 mm. gun, 1 L.M.G.
Ammunition carried:	100 rds. for gun, 2,400 rds. for M.G.
Dimensions:	Length, 13 ft. 2 in. Width, 6 ft. 1 in. Height, 7 ft. 6 in. Ground clearance, 1 ft. 2 in.
Performance:	Trench, 5 ft. 3 in. Step, 2 ft. 11 in. Water, 2 ft. 7 in. Gradient, 40°.
Engine:	83 h.p. 4 cyl. water cooled.
Speed:	Road, $12\frac{1}{2}$ m.p.h.; radius of action. 82 miles.
Suspension:	5 bogey wheels, 2 bogies, scissor articulation, 1 independent bogey wheel
Communication:	External, flag.

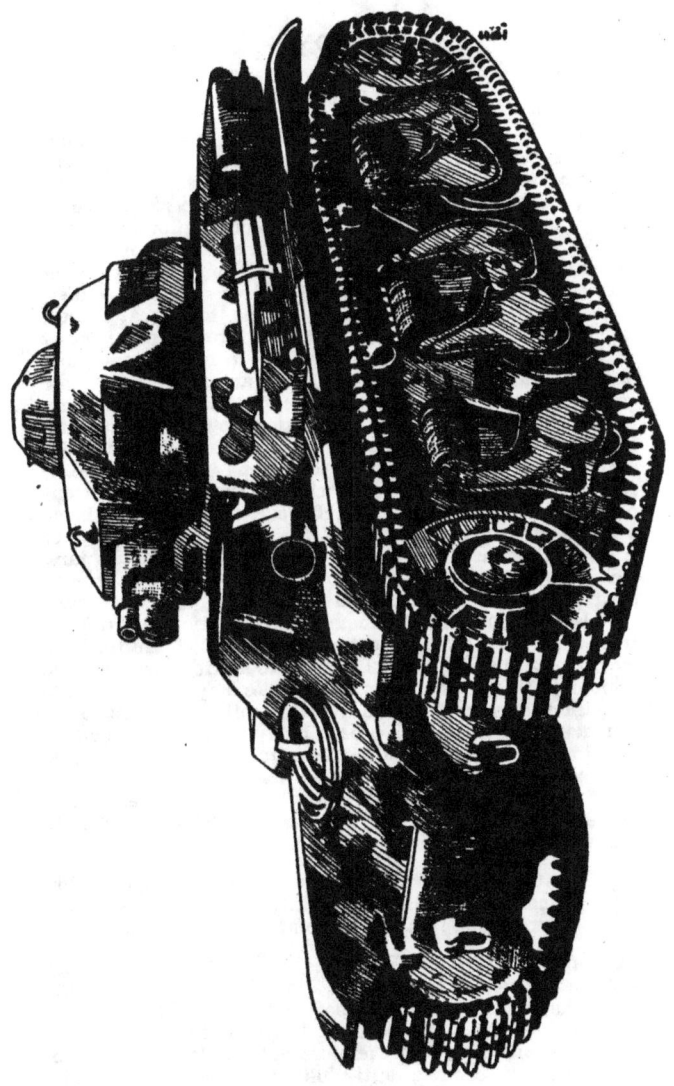

SECTION 5

VULNERABILITY OF A.F.Vs.

Before studying the diagram illustrating the armour thickness on A.F.Vs., it would be as well to consider the problems which the tank designers are continually aiming to overcome. Under ideal conditions the performance of anti-tank guns is superior to the protection afforded by the A.F.V. These ideal conditions, of course, are dependent on close range, a normal angle of impact and an anti-tank gunner with a good working knowledge of A.F.Vs. and their vulnerability. The first two conditions can be, to a great extent, offset by a tank designer in that thick armour plate set at angles makes it difficult for a gunner to register a normal impact. Naturally, the thickest plates are in the front of the vehicles, where they are most likely to have to stand up to enemy fire, and as the penetration powers of anti-tank guns fall off rapidly when an impact at an angle over 30° is registered, it is not difficult for a tank to be designed to give maximum protection by careful angling of the front plates and the turret plates. However, in order to achieve a practicable steering ratio and track pressure every ounce is conserved where possible, so that those parts of the tank which are less likely to come under fire carry thinner armour plating, such as the horizontal plates, back plates and even the plates on the sides of the hull are proportionately thinner than those front plates and turret plates which are the ones most likely to be shot at. Even so, no matter how thick the armour plating or how swift the vehicles may be, such points as vision slits and tracks remain uniformly vulnerable. Furthermore, A.F.Vs. have been lately proved to be vulnerable to armour piercing shells fired from aeroplane cannons. This will eventually lead to thicker armour plate being necessary on turret lids and power unit compartment. The vulnerability of enemy A.F.Vs. may, therefore, be shortly summarized as following :—

1. **Front plates.**—Proof against all except large calibre guns and an unreliable point of aim unless an angle approaching normal can be attained at impact.

2. **Turret.**—The above applies to the turret with the exception that the back and sides of the turret may have thinner armour plating than the front and the lid of the turret is most likely to be of quite thin armour plate and therefore vulnerable to shells from aircraft cannon, anti-tank grenades, S.T. grenades and made-up charges of explosive.

It is sometimes possible to jam the turret with S.A.A. or anti-tank rifle shots by firing at a point between the base of the turret and the top of the hull of the tank. In this way the traversing gear may become jammed. To offset this the Germans have either built up the front plates of the superstructure or welded plates shaped triangularly all along the front of the turret. The rear of the turret may also be protected in this way. The turret ring still remains vulnerable from the sides. The gun mantlet, although specially protected by extra armour plates, may still be vulnerable. A shot deforming the mantlet will prevent the gun from being elevated and depressed; when attacking a heavily armoured tank from the front the mantlet is the best target.

3. **Hull.**—Shots aimed behind the front sprocket or idle wheel and just in front of the rear sprocket or idle wheel, will put a tank out of action. In addition, in enemy A.F.Vs. in almost all cases a shot penetrating behind the front sprocket will kill the driver and a shot fairly low down in the centre of the tank stands a very good chance of shooting off the legs of the crew in the fighting cab. The sides of the tank are vulnerable and a good place to aim for, as they are in all cases vertical plates and a normal angle may not be so difficult to achieve. Also, in all enemy A.F.Vs. the power unit is situated at the rear as also are the fuel tanks, and a shot penetrating the engine compartment will render the tank immobile.

4. **Tracks.**—The tracks are most vulnerable to anti-tank mines, anti-tank grenades, S.T. grenades. They are not easy to hit with an anti-tank rifle or guns, and the former cannot always be relied upon to break the track, but, if the vehicle is approaching head-on and the only anti-tank weapon available is the anti-tank rifle, then it may be used against the tracks.

5. **Vision slits.**—If the tank is blinded the crew inside are also blinded, and it is impossible for them to fight their vehicles except in a most haphazard way. The tank may be blinded with incendiary grenades or smoke grenades. When blinded, an A.F.V. has few alternative courses of action : it either stops, possibly firing wildly, or turns round in an attempt to escape, or endeavours to rush through the screen at high speed in the hope of getting clear. Whatever course it takes, it should not be difficult for a determined soldier to put it out of action.

6. **Air louvres.**—Fumes and splinters will enter these openings. They are particularly vulnerable to incendiary grenades.

7. **Welded joints.**—In order to save weight, and for the sake of ease of production, the Germans use half-V weldings to join their armour plates. These are apt to break up under continuous fire.

8. It must be remembered that if a tank has had its tracks blown off by a mine or part of its suspension blown off by an anti-tank gun or other weapon, or had the power unit broken up, the crew inside may still be alive and as long as they are alive and their guns in working order they will be able to defend themselves with effect.

9. German tanks Pz. Kw. I, II, III and IV can be penetrated in all parts by the 2-pdr. anti-tank gun at ranges up to 500 yards.

10. An experienced tank commander will always endeavour to approach his objective obliquely, that is, with his guns firing over the left or right front corner of the tank. In this way he is least vulnerable to the anti-tank weapons ahead of him, who will find it extremely difficult to register a hit under 30° to normal impact. Anti-tank fire must, therefore, be held until a favourable shot is possible.

Pz. Kw. I. (Armour Plate)

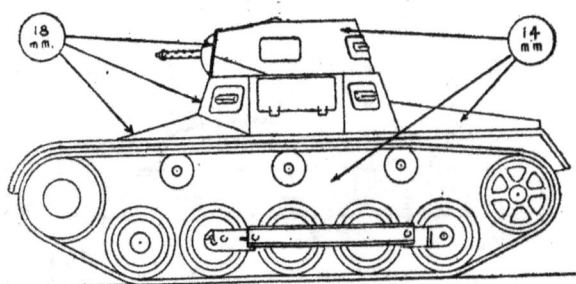

SIDE VIEW.

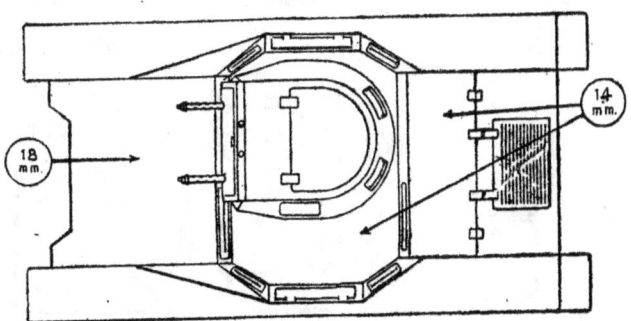

PLAN.

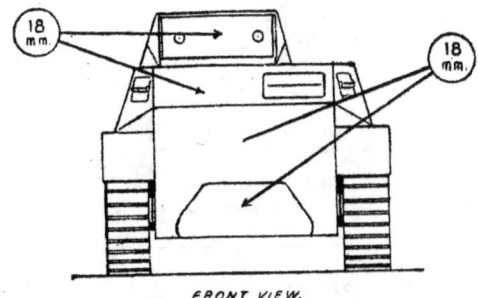

FRONT VIEW.

Pz. Kw. II.

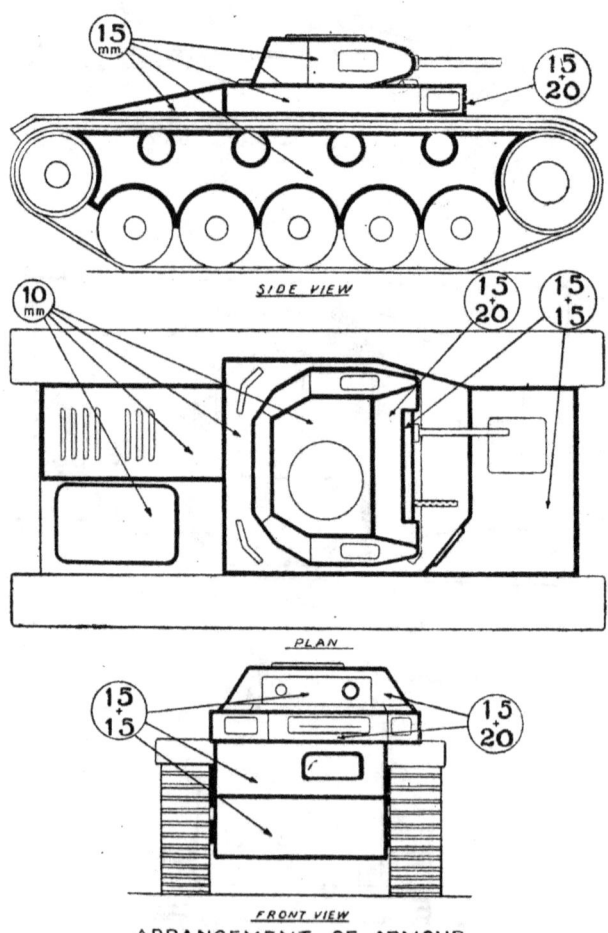

ARRANGEMENT OF ARMOUR

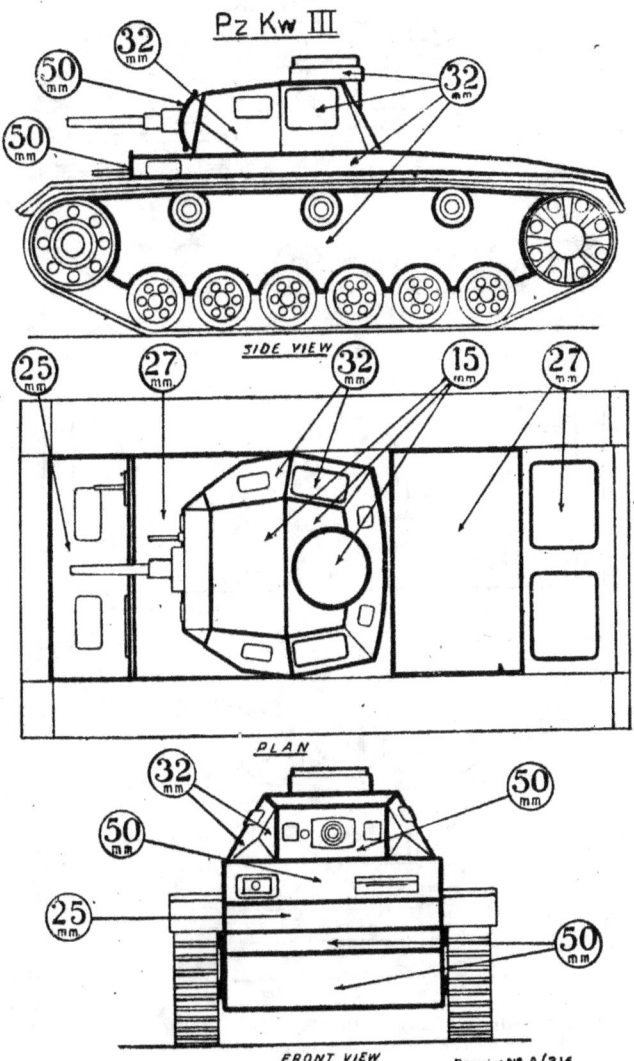

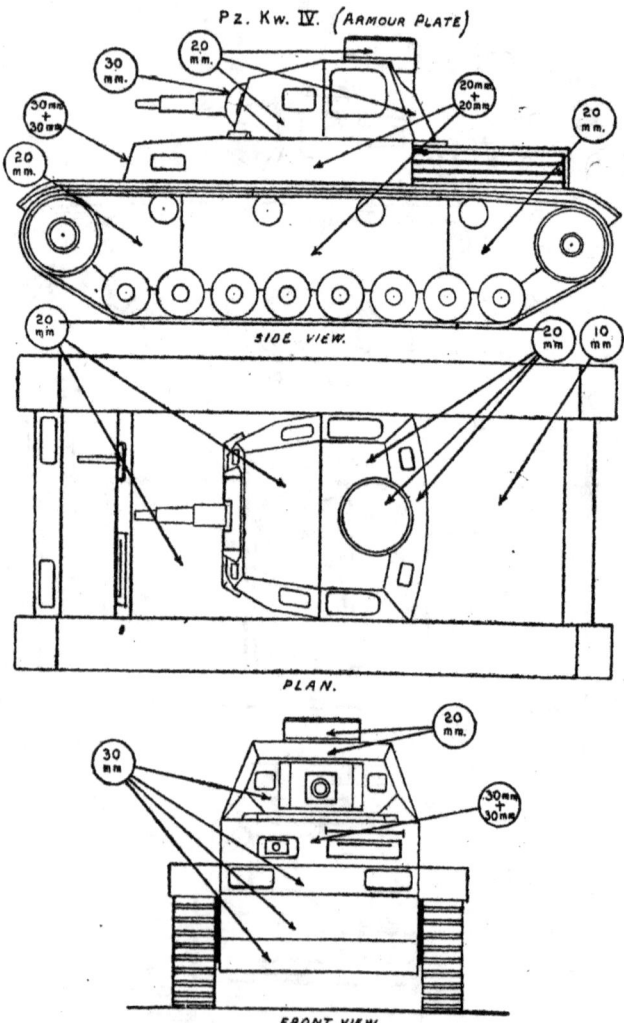

www.ingramcontent.com/pod-product-compliance
Lightning Source LLC
Chambersburg PA
CBHW050206130526
44591CB00035B/2296